AI-Driven Project Management

T.C. Catz
AI-Driven Project Management

All rights reserved
Copyright © 2024 by T.C. Catz

No part of this publication may be reproduced, distributed, or transmitted in any form or by any means, including photocopying, recording, or other electronic or mechanical methods, without the prior written permission of the publisher, except in the case of brief quotations embodied in critical reviews and certain other noncommercial uses permitted by copyright law.

Published by Spines
ISBN: 979-8-89569-681-1

AI-Driven Project Management

How the Rise of Artificial Intelligence is Changing the World

T.C. Catz

CONTENTS

Dedication

To all those who believe in the potential of artificial intelligence to make the world a better place, and who strive to ensure that this technology is developed and used responsibly, ethically, and with a deep respect for human values.

Why The N.E.R.D.Y. Way?

The **N.E.R.D.Y.** Way, a potent acronym that stands for k**N**owledge, **E**ducation, **R**esource, **D**iscovery for **Y**ou, embodies the spirit of this book. It's not just a journey through the world of AI; it's an invitation to embark on a lifelong adventure of learning, exploration, and constant evolution. As the field of AI progresses at an astonishing pace, so too must our understanding and engagement with it. The NERDY Way encourages you to embrace this dynamic and ever-changing landscape as a catalyst for personal growth and societal advancement.

Think of it as an ongoing dialogue, a conversation between you and the world of AI, where curiosity is your compass and exploration is your guide. This journey is not about reaching a destination; it's about the continuous process of learning, adapting, and evolving alongside the ever-expanding frontiers of AI. Embrace the challenges and opportunities that come with this journey, for within them lies the potential to unlock your own capabilities and contribute to a future where technology empowers humanity.

The N.E.R.D.Y. Way is a mindset, a philosophy that encourages you to approach AI with a sense of wonder and a spirit of inquiry. It's about recognizing the profound impact AI is having on every aspect of our lives and acknowledging its potential to reshape our world. This mindset fosters a deep appreciation for the transformative power of AI while also recognizing the critical need for responsible development and deployment. It's about understanding the intricate workings of AI systems, their strengths, and limitations, and using this knowledge to make informed decisions about their use.

The N.E.R.D.Y. Way isn't just about acquiring knowledge; it's about applying it to create a better future. This journey is about using your understanding of AI to solve global challenges, foster collaboration between humans and machines, and shape a future where technology serves as a force for good. It's about embracing the responsibility that comes with this knowledge, recognizing that AI's future depends on our collective efforts.

For those who choose to embark on this journey, the rewards are boundless. You will gain a deeper understanding of the world around you, develop valuable skills, and contribute to a future where technology serves as a force for good. It's an invitation to join the conversation, to contribute to the dialogue, and to shape the future of AI for the benefit of all. The N.E.R.D.Y. Way is a testament to the power of learning, collaboration, and continuous exploration, a journey that will enrich your life and help build a better future for everyone.

The N.E.R.D.Y. Way isn't just about understanding AI; it's about becoming a part of its evolution, a contributor to its progress, and a champion for its responsible development and deployment. It's a reminder that the future of AI is not a distant prospect; it's happening now, and it's up to us to shape it. It's a call to action, a reminder that we all have a role to play in this journey, and every step we take, every question we ask, every idea we share, helps us move closer to a brighter future.

Introduction

AI in Project Management: A Paradigm Shift

The integration of AI into project management represents a paradigm shift, moving from traditional methods to a more data-driven and intelligent approach. This transformation is reshaping the role of project managers, empowering them with new tools and capabilities to optimize projects and deliver exceptional outcomes.

This book explores the transformative impact of AI on project management, providing a comprehensive overview of AI concepts, its applications in engineering, IT and construction project management, and the potential benefits it offers. We will delve into the practical aspects of AI implementation, examining the tools, techniques, and best practices for leveraging AI to enhance project outcomes. We will also address the challenges associated with AI integration, including ethical considerations, data privacy, and the need for skilled professionals to manage and interpret AI-generated insights. Through real-world examples and case studies, this book empowers project managers, industry professionals, and students to embrace AI as a strategic tool for driving success in the ever-evolving landscape of project management.

WHAT IS ARTIFICIAL INTELLIGENCE

Artificial intelligence (AI) is rapidly transforming various industries, and project management is no exception. At its core, AI encompasses the ability of computer systems to mimic human intelligence, enabling them to learn, reason, and make decisions. This transformative technology is built upon a foundation of complex algorithms and vast datasets, allowing machines to analyze information, identify patterns, and generate insights that can inform human decision-making.

THE EVOLUTION OF AI: A HISTORICAL PERSPECTIVE

To fully grasp the impact of AI on project management, it is essential to investigate the history of this technology.

The seeds of AI were sown in the 1950s, when Alan Turing, a British mathematician and computer scientist, proposed the Turing Test, a test that was used to evaluate if a machine had the ability to act like a human. The same decade witnessed the birth of the field of artificial intelligence as a distinct discipline, with the Dartmouth Summer Research Project on Artificial Intelligence laying the foundation for its future development.

The 1960s and 1970s marked a period of intense research and exploration in AI, focusing on developing expert systems, rule-based programs designed to emulate the reasoning abilities of human experts in specific domains. These early systems, while limited in their scope, demonstrated the potential of AI to solve complex problems and provide valuable assistance to human practitioners.

The 1980s saw the rise of machine learning, with the development of algorithms capable of learning from data without explicit programming. This breakthrough opened up new possibilities for AI, expanding its applications beyond rule-based systems to encompass data-driven approaches to solving problems.

The late 20th century witnessed a confluence of factors that accelerated AI's progress, including the exponential growth of computing power, the availability of massive datasets, and the development of new algorithms. This period saw the emergence of deep learning, a powerful subfield of machine learning characterized by the use of artificial neural networks.

The 21st century has been marked by a surge in AI advancements, fueled by the increasing availability of data, the development of cloud computing platforms, and the rise of mobile devices. AI has become ubiquitous, permeating various aspects of our lives, from personalized recommendations on streaming services to self-driving cars.

Key Concepts and Terminology

This section delves into the fundamental concepts and terminology that are essential for understanding AI in project management. It's like learning the alphabet before you can read a book.

Algorithms: The Brains of AI

Imagine a recipe, but instead of making cookies, it tells a computer how to solve a problem. That's essentially what an algorithm is. It's a set of instructions that a computer follows to perform a specific task. In AI, algorithms are the core of how machines learn and make decisions.

Think of it like this:

Recipe for a Cake: You follow steps like mixing ingredients, baking, and frosting. The recipe is the algorithm.

AI Algorithm: The algorithm might analyze data like historical project timelines, resource usage, or market trends. Based on these data points, it can generate predictions for future projects, identify potential risks, or even recommend strategies for optimizing resource allocation.

Data Sets: Feeding the AI Machine

Data is the fuel that powers AI. A data set is a collection of information used to train AI models. It's like a library of examples that the AI algorithm uses to learn patterns and make predictions.

Think of it like this:

Learning to Recognize a Cat: A child learns to identify a cat by seeing many pictures of cats. The pictures are the data set.

AI Learning to Predict Project Delays: An AI model can learn to predict project delays by analyzing data sets of past projects, including factors like weather conditions, resource availability, and past performance.

Neural Networks: Mimicking the Brain

Neural networks are a type of AI algorithm inspired by the structure of the human brain. They are composed of interconnected nodes (neurons) that process information in a layered structure. Each layer learns different features of the data, and by working together, they can solve complex problems.

Think of it like this:

Brain Cells: Imagine your brain as a network of interconnected neurons, each responsible for processing information in a specific way.

Neural Network: AI neural networks mimic this structure, using interconnected nodes to learn patterns from data. They can be used for various tasks, like image recognition, language translation, and even to forecast project outcomes.

Machine Learning: AI that Learns from Experience

One of the key pillars of AI is machine learning, which empowers systems to learn from data without explicit programming. This process involves training algorithms on large datasets, allowing them to identify patterns, make predictions, and improve their accuracy over time. Imagine a machine learning algorithm tasked with classifying images of cats and dogs. By being exposed to countless examples of each animal, the algorithm gradually learns to distinguish between them based on visual cues, such as fur patterns, ear shapes, and facial features. It's like teaching a child to ride a bike by observing them and providing feedback. The machine learning algorithm learns by analyzing data sets and identifying patterns.

Think of it like this:

Child Learning to Ride a Bike: A child learns by trial and error, with feedback from parents.

Machine Learning for Resource Allocation: An AI model can learn to optimize resource allocation by analyzing data on past projects, considering factors like skill sets, availability, and project requirements.

Deep Learning: A Powerful Tool for Complex Data

Deep learning is a subset of machine learning that uses deep neural networks, with multiple layers of interconnected nodes, to learn complex patterns from data. It's particularly powerful for tasks involving large amounts of data, like analyzing images, understanding natural language, or predicting complex events. Deep learning is characterized by the use of artificial neural networks, inspired by the structure and function of the human brain. These networks consist of interconnected nodes, or neurons, organized in layers. Each connection between neurons has a weight associated with it, which is adjusted during the learning process to optimize the network's performance. Deep learning excels at tasks involving complex patterns and large amounts of data, such as image recognition, natural language processing, and speech synthesis.

Think of it like this:

Recognizing a Face: Your brain processes a face in a complex way, using multiple layers of neurons to identify features like eyes, nose, and mouth.

Deep Learning for Image Recognition: A deep learning algorithm can recognize images of people, objects, and even specific events by analyzing the images with multiple layers of neural networks.

Natural Language Processing (NLP): AI that Understands Language

NLP is a branch of AI that focuses on enabling computers to understand and interact with human language. It involves algorithms that can analyze text data, extract meaning, and generate text in a human-like way. NLP enables machines to understand, interpret, and generate human language. NLP algorithms allow computers to process text and speech, extracting meaning, identifying sentiment, and translating languages. Imagine a chatbot designed to assist customers with product inquiries. NLP empowers this chatbot to understand customer queries, provide relevant responses, and even engage in natural-sounding conversations.

Think of it like this:

Reading a Book: You understand the meaning of a book by processing the words and sentences.

NLP for Analyzing Project Documents: NLP algorithms can analyze project documents, identify key information, and even translate text from one language to another.

These fundamental concepts are essential for understanding how AI works and its potential applications in project management. In the upcoming chapters, we will delve deeper into these concepts, explore their specific applications in engineering and construction project management, and uncover the transformative potential of AI for optimizing projects and driving innovation across these industries.

Types of Artificial Intelligence

To understand how AI can benefit project management, we must first grasp the different types of AI and their capabilities. AI can be broadly categorized into three primary types: narrow AI, general AI, and super AI.

Narrow AI: The AI We Use Everyday

Narrow AI is the most prevalent type of AI in use today. It excels at performing specific tasks, demonstrating intelligence in a limited domain. Imagine a chess-playing program that can defeat even the best human players. It may be a master of strategy and tactics within the confines of a chessboard, but it wouldn't be able to write a poem, compose a symphony, or understand a complex legal document. Narrow AI is designed to be specialized, excelling in a specific area.

Here are some everyday examples of narrow AI at work:

- **Spam filters:** These intelligent programs analyze emails for patterns and keywords to identify and filter out spam messages.
- **Recommendation engines:** These algorithms analyze your past preferences and browsing history to suggest products, movies, or articles you might enjoy.
- **Virtual assistants:** From Siri to Alexa, virtual assistants use natural language processing to understand your voice commands and respond appropriately.
- **Fraud detection systems:** AI algorithms monitor transactions for suspicious patterns and anomalies, helping to prevent fraud.

- **Medical diagnosis:** AI systems can analyze medical images, identify patterns, and assist doctors in making diagnoses.

While narrow AI may not be able to perform tasks outside its designated domain, it already plays a significant role in our lives. It automates tasks, enhances efficiency, and provides personalized experiences.

General AI: The Quest for Human-Level Intelligence

General AI represents a different level of intelligence. It aims to achieve human-level cognitive abilities, encompassing a wide range of tasks and domains. Imagine an AI that can reason, learn, problem-solve, and understand complex concepts just as humans do. It could potentially perform any task a human can, from writing a novel to designing a building to conducting scientific research.

General AI is still a work in progress. While we've made significant strides in developing AI that can excel in specific areas, creating AI that can truly think and reason like humans remains a formidable challenge. The complexity of human cognition, with its ability to learn, adapt, and understand emotions, is a significant hurdle.

Super AI: The Future of Intelligence

Super AI represents the hypothetical future of AI. It surpasses human intelligence in all aspects, exceeding our cognitive abilities in every way. Imagine an AI that can solve complex problems, innovate at an unprecedented pace, and reshape our world in unimaginable ways.

Super AI is a fascinating but distant prospect. It raises profound ethical questions about the relationship between humans and AI, the potential for unintended consequences, and the need for responsible development. While super AI may seem like science fiction, it's an area of intense research and speculation.

EXAMPLES AND APPLICATIONS

The world around us is rapidly becoming infused with AI, and its influence extends far beyond the realm of science fiction. AI is no longer a futuristic concept; it's a tangible force shaping our daily lives and revolutionizing industries across the globe. From the healthcare sector to the financial markets, AI is demonstrating its immense potential to transform how we live, work, and interact with the world.

In healthcare, AI is making a profound impact on patient care and diagnostics. Imagine a system that can analyze medical images, such as X-rays and MRIs, with unparalleled accuracy, identifying potential abnormalities that human eyes might miss. AI-powered diagnostic tools are already being used to detect diseases like cancer at earlier stages, when treatment is most effective. Moreover, AI is assisting doctors in creating personalized treatment plans tailored to individual patients' needs and medical histories.

Beyond diagnostics, AI is also improving patient care by automating administrative tasks, allowing medical professionals to dedicate more time to their patients. AI-powered chatbots can answer patient questions, schedule appointments, and provide basic health information, freeing up doctors and nurses for more complex tasks.

In finance, AI is revolutionizing how we manage our money and make investment decisions. AI-powered algorithms are analyzing vast amounts of financial data in real time, identifying patterns and trends that humans might miss. This data-driven approach is enabling investment managers to make more informed decisions, potentially leading to higher returns and reduced risk.

Moreover, AI is transforming fraud detection and risk management in the financial industry. By identifying suspicious transactions and patterns, AI systems can help banks and financial institutions prevent fraud and safeguard customer accounts.

The impact of AI extends to the realm of customer service as well. AI-powered chatbots and virtual assistants are becoming increasingly prevalent, providing instant support to customers across various industries. These AI systems can answer frequently asked questions, resolve common issues, and even recommend products or services tailored to individual customer preferences.

The automotive industry is also undergoing a significant AI-powered transformation. Self-driving cars, powered by AI algorithms that can navigate complex environments, are becoming a reality. These vehicles promise to revolutionize transportation, increasing safety, reducing traffic congestion, and offering greater accessibility for those who are unable to drive.

AI is even making its way into the world of education, revolutionizing how we learn and teach. AI-powered tutors can provide personalized learning experiences, adapting to individual student needs and learning styles. These intelligent tutors can identify areas where students need extra help, offer targeted instruction, and track progress in real time.

The influence of AI on the legal profession is growing as well. AI-powered systems can analyze legal documents, identify relevant precedents, and assist lawyers in preparing legal arguments, saving time and effort. Moreover, AI can be used to predict legal outcomes, providing valuable insights to clients and attorneys.

The applications of AI are truly boundless, and its transformative potential continues to expand. As AI technologies advance, we can expect to see even more groundbreaking applications emerge across diverse industries, shaping the future of our world in profound and exciting ways.

AI's Transformative Power: Reshaping Project Management

The impact of AI on project management is already evident across various industries. From engineering to construction, AI-powered tools are being leveraged to optimize project planning, scheduling, resource allocation, risk management, and decision-making.

In **engineering projects**, AI assists with design optimization, structural analysis, and material selection. Machine learning algorithms can analyze vast datasets of past projects, identifying optimal design parameters, minimizing material usage, and predicting potential failures. AI-powered simulations can accelerate the design process, allowing engineers to explore multiple design scenarios and identify the most efficient and cost-effective solutions.

In the **construction industry**, AI is revolutionizing project planning and scheduling. AI-powered tools can analyze historical data, identify potential delays, and optimize resource allocation, ensuring projects are completed on time and within budget. AI-driven systems are also being used for cost estimation, risk management, and quality control, improving the accuracy and efficiency of these critical aspects of construction projects.

AI's ability to process vast amounts of data, identify patterns, and make predictions is revolutionizing project management. It provides project managers with powerful tools to optimize project planning, resource allocation, risk management, and decision-making. However, AI's impact extends far beyond simple automation. It offers the potential to transform project management into a more data-driven, intelligent, and efficient process.

AI's Impact on Project Management: Key Benefits

AI offers a myriad of benefits for project management, including:

- **Enhanced Predictive Analytics:** AI-powered tools can analyze historical data to identify patterns and trends, allowing project managers to make more accurate forecasts about project timelines, resource requirements, and potential risks.
- **Improved Automation:** Repetitive tasks, such as data entry, scheduling, and reporting, can be automated

using AI, freeing up valuable time for project managers to focus on strategic activities.

- **Data-Driven Decision-Making:** AI provides project managers with data-driven insights, supporting informed decisions about resource allocation, risk mitigation, and project prioritization.

With its capabilities in data analysis, pattern recognition, and automation, AI is poised to revolutionize project management. Imagine AI tools that can:

- **Enhanced Collaboration and Communication:** AI-powered tools can facilitate seamless communication and collaboration between project stakeholders, improving information sharing and transparency.
- **Increased Efficiency and Productivity:** By automating tasks, optimizing resource allocation, and providing data-driven insights, AI can significantly enhance project efficiency and productivity.
- **Forecast project timelines:** Analyze historical project data to predict project durations, identify potential delays, and optimize scheduling.
- **Identify and assess risks:** Use data analysis to identify potential risks, assess their likelihood and impact, and recommend mitigation strategies.
- **Optimize resource allocation:** Allocate resources effectively based on project demands, skill sets, and availability, ensuring optimal resource utilization.
- **Automate repetitive tasks:** Free up project managers' time by automating tasks such as data entry, scheduling, and report generation.

- **Provide data-driven insights:** Analyze project data to provide valuable insights, support decision-making, and improve project outcomes.

AI can enhance efficiency, improve productivity, and empower project managers to make more informed decisions. It can transform project management from a reactive approach to a proactive, data-driven process.

THE POTENTIAL OF AI FOR PROJECT MANAGEMENT

The realm of project management is undergoing a profound transformation, driven by the rapid advancements in artificial intelligence (AI). AI is no longer a futuristic concept; it's a powerful tool that's reshaping the way projects are planned, executed, and delivered across industries.

Imagine a project manager equipped with an AI-powered assistant that can analyze vast amounts of data, predict potential risks, and suggest optimal resource allocation strategies. This is the reality that AI is bringing to project management. AI-driven project management empowers project leaders to make more informed decisions, automate repetitive tasks, and unlock new levels of efficiency.

- **The Power of AI-Driven Insights:** AI's ability to analyze data and extract valuable insights is revolutionizing project management. Project managers can leverage AI to:
- **Predict Project Timelines and Deadlines:** AI algorithms can analyze historical data from previous

projects, market trends, and resource availability to predict project timelines with greater accuracy. This allows for proactive planning, resource allocation, and risk mitigation to prevent delays and ensure project delivery within the set timeframe.

- **Identify and Assess Project Risks:** AI algorithms can analyze vast amounts of data to identify potential risks that could impact project success. This includes factors like supplier reliability, market fluctuations, and unforeseen technological challenges. By identifying risks early on, project managers can develop mitigation strategies and take proactive steps to minimize their impact.

- **Optimize Resource Allocation:** AI can analyze project requirements, skill sets, and availability of resources to optimize allocation. It can recommend the most efficient deployment of personnel, equipment, and financial resources, ensuring that the right resources are available at the right time.

- **Enhance Cost Estimation and Budgeting:** AI can analyze historical data, market trends, and project specifications to provide accurate cost estimations. It can identify potential cost overruns and suggest cost-saving measures, leading to more effective budgeting and financial management.

- **Automating Tasks for Increased Efficiency:** AI-powered tools can automate a wide range of repetitive tasks, freeing up valuable time for project managers to focus on strategic decision-making and complex problem-solving. Here are some examples:

- **Data Entry and Reporting:** AI can automate data entry from various sources, such as spreadsheets, emails, and project management systems. This eliminates manual data entry errors and allows for quick and accurate reporting.

- **Scheduling and Task Management:** AI-powered tools can automate task scheduling, resource allocation, and progress tracking. They can alert project managers to potential delays, suggest alternative solutions, and provide real-time updates on project status.

- **Document Generation and Review:** AI can automate the generation of project documents, such as reports, proposals, and presentations. It can also review documents for accuracy and consistency, reducing the need for manual proofreading.

- **Transforming Collaboration and Communication:** AI is transforming the way project teams communicate and collaborate, breaking down silos and fostering seamless information sharing. Here are some key benefits:

- **Real-Time Information Sharing:** AI-powered platforms can facilitate real-time information sharing, ensuring that all stakeholders are informed about project progress, changes, and updates. This transparency fosters trust and accountability.

- **Improved Communication Flow:** AI-driven tools can analyze communication patterns and identify areas for improvement. They can suggest better communication channels, optimize meeting schedules, and streamline information flow within project teams.

- **AI-Powered Collaboration Tools:** AI can power collaborative tools like chatbots, virtual assistants, and automated meeting scheduling. These tools facilitate seamless communication, task delegation, and problem-solving within project teams.

- **Ethical Considerations and Future Outlook:** While AI offers tremendous potential for project management, it's essential to address ethical considerations, such as data privacy, bias, and responsible AI development. Organizations must implement robust data security measures, ensure fair and unbiased AI algorithms, and prioritize ethical AI practices. AI is not just a technological advancement; it's a cultural shift. As we embrace AI in project management, we must also be mindful of its ethical implications. We need to ensure that AI systems are developed and used responsibly, avoiding bias, ensuring data privacy, and promoting transparency.

CHAPTER 1

AI in Engineering Project Management

AI can be a powerful tool for optimizing design and engineering processes, streamlining material selection, and enhancing structural analysis in engineering projects. By harnessing the capabilities of machine learning and deep learning algorithms, AI can analyze vast amounts of data, identify patterns, and generate insights that human engineers may miss.

AI-Powered Design Optimization

One of the most transformative applications of AI in engineering is design optimization. Traditionally, engineers have relied on their expertise and intuition to create designs that meet specific criteria. However, AI-powered design optimization tools can automate this process, exploring a much wider range of design possibilities and identifying optimal solutions that maximize performance, minimize costs, and optimize material usage.

Let's consider a real-world example. Imagine a civil engineer designing a bridge. Using AI-powered design optimization software, the engineer can input various parameters, such as load capacity, material properties, and environmental conditions. The AI algorithm then explores countless design variations, considering factors like beam shape, truss configuration, and support placement. The software analyzes each design based on its strength, stability, and cost-effectiveness, ultimately presenting the engineer with a selection of highly optimized bridge designs.

This process not only saves time and effort for the engineer but also opens the door to innovative and highly efficient designs that might not have been conceived using traditional methods.

AI-Driven Structural Analysis

Structural analysis is another critical area where AI can significantly contribute. AI algorithms can analyze complex structures and identify potential weaknesses, stress points, and areas of failure. This information allows engineers to make informed decisions regarding material selection, reinforcement, and design modifications to ensure the structural integrity and safety of buildings, bridges, and other infrastructure.

Consider the construction of a high-rise building. Using AI-powered structural analysis tools, engineers can simulate various load conditions, wind pressures, and seismic events. The AI algorithm analyzes the stresses and strains on different structural elements, providing insights into their behavior and identifying areas that may require reinforcement or design adjustments.

This proactive approach to structural analysis can help prevent catastrophic failures, ensure the safety of occupants, and reduce the risk of costly repairs or rebuilds.

AI-Assisted Material Selection

Material selection is a crucial decision in any engineering project. AI can revolutionize this process by providing data-driven insights into the properties, cost, and availability of various materials. By analyzing historical data on material performance, environmental factors, and construction trends, AI can assist engineers in choosing the most suitable materials for a specific project.

For example, consider a project involving a high-temperature application. AI can analyze the performance of various heat-resistant materials under specific temperature conditions. The AI algorithm can then recommend materials that best balance performance, durability, and cost-effectiveness based on the project's requirements. This approach not only optimizes material selection but also helps engineers make informed choices that minimize environmental impact and maximize resource utilization.

Beyond Design: AI's Impact on Engineering Operations

The applications of AI in engineering extend far beyond design. AI-powered tools can also be used for:

- **Predictive maintenance:** AI can analyze sensor data from equipment to predict potential failures before they occur, enabling proactive maintenance and reducing downtime.
- **Construction robotics:** AI-powered robots are increasingly used for tasks like welding, bricklaying,

and concrete pouring, improving efficiency and safety
in construction.

- **Construction Safety management:** AI can analyze
real-time data from construction sites to identify
potential hazards and ensure compliance with safety
protocols.

PREDICTIVE MAINTENANCE AND ASSET MANAGEMENT

Imagine a world where equipment failures are predicted before
they occur, maintenance schedules are optimized based on real-
time data, and assets live longer than ever before. This is the
promise of AI in predictive maintenance and asset management,
a revolution sweeping across engineering and construction
industries.

At the heart of this revolution is the ability of AI to learn from
vast amounts of data, identifying patterns and anomalies that
signal impending equipment failure. By analyzing sensor readings, operational logs, and historical maintenance records, AI
algorithms can forecast potential breakdowns with remarkable
accuracy.

Take the example of a construction company managing a fleet of
heavy-duty machinery. AI can analyze data from the machinery's
sensors – such as engine temperature, vibration levels, and fuel
consumption – to detect subtle changes that indicate a potential
failure. This allows the company to proactively schedule maintenance before the equipment breaks down, avoiding costly downtime and repairs.

But the benefits extend beyond simply predicting failures. AI can also optimize maintenance schedules, ensuring that maintenance is performed only when needed, minimizing downtime and maximizing asset uptime. Imagine a scenario where AI analyzes the performance of a data center's cooling system, identifying potential issues before they become critical. This allows IT professionals to schedule maintenance during off-peak hours, preventing disruptions to critical operations.

Moreover, AI can play a pivotal role in extending the lifespan of assets. By analyzing operational data, AI can identify factors that contribute to wear and tear, allowing engineers to implement preventive measures to mitigate these factors. This can include optimizing operating parameters, adjusting maintenance schedules, and even suggesting upgrades that extend the asset's useful life.

For instance, in the construction industry, AI can be used to monitor the performance of cranes and other heavy equipment, identifying potential problems before they cause damage. By analyzing data from sensor readings and operational logs, AI can recommend adjustments to operating parameters, reducing stress on the equipment and extending its lifespan.

Beyond these individual applications, AI can be integrated into broader asset management strategies. For example, AI can analyze data from various sources to identify potential bottlenecks in the supply chain, optimizing inventory levels and reducing the risk of delays.

Consider a scenario where an engineering firm uses AI to analyze the performance of its project management system. The AI can identify patterns in project delays, resource allocation issues, and communication bottlenecks. These insights enable the company to implement improvements to its project management processes, leading to increased efficiency and productivity.

However, the successful implementation of AI in predictive maintenance and asset management requires careful consideration of several factors.

Data Quality and Availability

At the core of AI lies data. Without high-quality, readily available data, AI algorithms cannot generate accurate predictions or insights. Therefore, organizations must invest in data collection, storage, and management systems that ensure data quality and accessibility.

Algorithm Selection

Selecting the right AI algorithms is crucial for achieving desired outcomes. Different AI models are suited to specific types of data and tasks. Organizations need to choose algorithms that are well-suited to their specific needs and data sets.

Integration with Existing Systems

Integrating AI solutions into existing asset management systems can be challenging. It requires careful planning and coordination to ensure smooth data flow and integration between different systems.

Skill Development

The successful implementation and management of AI-powered solutions require a workforce with specialized skills in data science, machine learning, and AI applications. Organizations need to invest in training and development programs to equip their staff with the necessary expertise.

Ethical Considerations

As AI plays a more prominent role in asset management, it's essential to consider ethical implications. This includes ensuring that AI algorithms are unbiased, transparent, and explainable. Organizations need to develop robust ethical frameworks for the responsible use of AI in asset management.

The Future of AI in Asset Management

The future of AI in asset management is bright. As AI technology continues to evolve, we can expect even more sophisticated applications that further enhance efficiency, safety, and sustainability in engineering, IT, and construction.

Here are some key trends shaping the future of AI in asset management:

Edge Computing: AI algorithms will be deployed at the edge of the network, closer to data sources, enabling real-time analysis and faster decision-making.

Internet of Things (IoT): The proliferation of connected devices will generate even larger volumes of data, providing richer insights for AI-powered asset management.

Digital Twins: Virtual representations of physical assets will enable more comprehensive and accurate analysis, leading to improved decision-making.

Predictive Maintenance Optimization: AI algorithms will become more sophisticated in predicting equipment failures, leading to more accurate and efficient maintenance schedules.

Sustainable Asset Management: AI will play a crucial role in optimizing asset utilization, reducing waste, and promoting sustainability in engineering, IT, and construction projects.

AI is already transforming asset management, driving efficiency, safety, and sustainability across industries. As AI technology continues to evolve, its role in asset management will only become more significant, empowering organizations to optimize their assets, maximize uptime, and achieve greater project success.

CONSTRUCTION ROBOTICS

The construction industry is undergoing a radical transformation, with AI-powered robots and automation taking center stage. These technological advancements are revolutionizing traditional construction practices, offering unprecedented levels of efficiency, precision, and safety. Imagine a future where robots seamlessly perform complex tasks like welding, bricklaying, and even painting, while AI algorithms optimize construction schedules, predict potential risks, and ensure quality control. This future is no longer a distant dream; it's becoming a reality, driven by the power of AI.

One of the most prominent areas where AI is making its mark in construction is robotics. AI-powered robots are being deployed on construction sites, performing tasks that were once considered too dangerous, repetitive, or time-consuming for human workers. These robots can operate with incredible accuracy and precision, minimizing human error and improving overall project efficiency.

Welding is a prime example of a task that is being revolutionized by AI-powered robots. Traditional welding methods often require skilled human welders, who can be prone to fatigue and errors. AI-powered welding robots, however, can work tirelessly without breaks, delivering consistent and high-quality welds. These robots use advanced sensors and algorithms to precisely control the welding torch, ensuring accurate positioning and heat application. This not only enhances the quality of welds but also significantly reduces the risk of accidents and injuries associated with manual welding.

Bricklaying is another construction task that is being transformed by AI-powered robots. These robots are equipped with sophisticated vision systems and robotic arms that can lay bricks with astonishing speed and accuracy. AI algorithms analyze the building plans and determine the optimal brick placement, minimizing waste and ensuring precise alignment. This results in faster construction times, reduced labor costs, and improved structural integrity of buildings.

The use of AI-powered robots in construction goes beyond just welding and bricklaying. Robots are being developed to perform a wide range of tasks, including:

Concrete pouring: AI-powered robots can precisely pour concrete, ensuring consistent thickness and minimizing the risk of spills or uneven surfaces.

Rebar installation: Robots can install rebar reinforcement in concrete structures with high precision and accuracy, reducing the risk of errors and ensuring structural stability.

Painting: AI-powered robots can paint walls and other surfaces with incredible precision and speed, eliminating the need for manual labor and ensuring consistent paint application.

These are just a few examples of how AI-powered robots are transforming construction practices. As AI technology continues to advance, we can expect to see even more sophisticated and capable robots being deployed on construction sites. These robots will not only perform complex tasks with greater accuracy and efficiency, but they will also work collaboratively with human workers, creating a safer and more productive work environment.

Construction Safety Management

The construction industry, notorious for its inherent risks and safety concerns, stands to benefit significantly from the integration of AI. AI is playing a critical role in construction safety. AI-powered systems can monitor construction sites in real-time, identifying potential hazards and alerting workers to potential risks. These systems can also track worker movements, identify unsafe practices, and ensure compliance with safety regulations. This helps to reduce accidents, injuries, and fatalities on construction sites.

AI-powered systems can analyze massive datasets, identifying patterns and trends that may not be readily apparent to human observers. This allows for proactive safety measures and a more comprehensive understanding of potential hazards. One of the most prominent applications of AI in safety management is the development of predictive models. These models can analyze historical data on accidents, near misses, and other safety incidents to identify patterns and predict future occurrences. For example, AI can analyze data on weather conditions, worker fatigue, and equipment malfunction to predict the likelihood of accidents on construction sites. This information can then be used to implement preventative measures, such as adjusting work schedules, providing additional training, or implementing stricter safety protocols.

Beyond predictive modeling, AI can also play a crucial role in real-time hazard detection. Cameras and sensors equipped with AI algorithms can monitor construction sites for potential dangers, such as falling objects, unsafe scaffolding, and hazardous materials. These systems can immediately alert workers and site managers to potential threats, allowing for swift action to mitigate the risk. For instance, AI-powered cameras can be used to monitor workers wearing safety equipment and alert supervisors if someone is not complying with safety regulations.

Furthermore, AI can enhance risk assessment and mitigation by analyzing complex data sets that traditional methods may struggle to handle. This includes data on soil conditions, geological factors, environmental impacts, and construction materials. By leveraging machine learning and deep learning techniques, AI can identify subtle risks that could be overlooked by human engineers. This can lead to more effective and efficient risk mitigation strategies, reducing the likelihood of costly delays, accidents, and project failures.

Imagine a scenario where AI is used to analyze a large-scale construction project, incorporating data from various sources, including weather forecasts, soil analysis, and historical data on similar projects. The AI system can identify potential risks, such as the likelihood of landslides, flooding, or material defects. Based on this analysis, the system can recommend appropriate mitigation strategies, such as reinforcing slopes, adjusting construction schedules, or procuring higher-quality materials. This proactive approach can significantly enhance the safety of the project and minimize the potential for accidents and delays.

Another significant benefit of AI in safety management is its ability to improve worker training and education. AI-powered training programs can personalize the learning experience for each individual, focusing on their specific needs and areas for improvement. This can lead to a more effective and engaging training experience, resulting in a workforce that is better equipped to handle safety risks. For example, AI can analyze individual worker performance data, identifying areas where they may need additional training or support. Based on this analysis, AI can create personalized training programs that focus on specific safety protocols, hazard identification, and emergency procedures.

AI can also facilitate the development of comprehensive safety manuals and guidelines. By analyzing a vast library of safety documents, regulatory codes, and industry best practices, AI can generate comprehensive and tailored safety manuals for specific construction projects. These manuals can incorporate the latest safety regulations, highlight potential risks specific to the project, and provide detailed instructions for handling hazardous materials and equipment.

However, it's essential to acknowledge that the successful integration of AI in safety management requires a careful consideration of ethical and legal implications. Concerns regarding data privacy, algorithmic bias, and accountability must be addressed. Transparency and explainability of AI-driven decision-making are critical to building trust and ensuring responsible use of this technology.

In conclusion, AI holds immense potential for revolutionizing safety management in the construction industry. By leveraging its capabilities for predictive modeling, real-time hazard detection, risk assessment, worker training, and the development of comprehensive safety guidelines, AI can significantly reduce the risk of accidents, injuries, and project failures. However, responsible and ethical implementation is crucial to ensure the safe and effective use of AI for enhancing safety on construction sites. The construction industry is now at a pivotal moment, where AI can transform its approach to safety and usher in a new era of enhanced security and worker well-being.

Case Studies: Real World Examples of AI in Engineering

The transformative potential of AI in engineering is best understood through real-world examples. Let's delve into a few compelling case studies that showcase the successful implementation of AI, highlighting both the benefits and challenges encountered.

1. AI-Powered Predictive Maintenance in Wind Turbine Operations:

Imagine vast wind farms stretching across the landscape, generating clean energy. However, these turbines are complex machines, susceptible to wear and tear, and potential failures can disrupt power generation and incur significant costs. Enter AI, which can revolutionize maintenance practices.

- **Case Study:** A leading wind turbine manufacturer partnered with an AI solutions provider to develop a predictive maintenance system. The system utilizes data from sensors embedded in the turbines, including vibration patterns, temperature fluctuations, and operational parameters. Using machine learning algorithms, the AI system analyzes this data to predict potential failures, often weeks or even months in advance.
- **Benefits:** This AI-powered solution has drastically reduced downtime for turbines. By predicting failures before they occur, maintenance teams can proactively schedule repairs, minimizing disruptions to power generation. This translates into significant cost savings, improved reliability, and increased energy output from the wind farm.
- **Challenges:** While the benefits are undeniable, implementing such an AI system requires careful consideration. The initial investment in sensors and AI software can be substantial. Moreover, training the AI system requires a large dataset of historical turbine data, which can be challenging to collect and manage.

2. AI-Driven Design Optimization for Bridge Construction:

Bridging the gap between complex engineering challenges and innovative design solutions is where AI steps in. The design and construction of bridges is a meticulous process, demanding accuracy, efficiency, and structural integrity. AI is now playing a vital role in optimizing this process.

- **Case Study:** A renowned engineering firm tasked with designing a new bridge utilized AI-powered software for design optimization. The software analyzed multiple design parameters, including material properties, load distribution, and environmental factors. By iterating through thousands of potential designs, the AI system identified the most efficient and structurally sound bridge design, minimizing material usage and construction costs.
- **Benefits:** The AI-driven design optimization process significantly reduced the time and effort required for bridge design. It also led to a lighter and more cost-effective bridge design while ensuring optimal structural integrity. This demonstrates AI's ability to generate innovative and sustainable solutions for complex engineering projects.
- **Challenges:** Implementing AI for design optimization requires expertise in both structural engineering and AI algorithms. Training the AI system with relevant data and validating its outputs are essential steps to ensure the accuracy and reliability of the design solutions generated.

3. AI-Assisted Risk Management in Construction Projects:

Construction projects are inherently complex, involving numerous stakeholders, diverse environmental factors, and potential risks. AI is emerging as a powerful tool for mitigating risks and ensuring project success.

- **Case Study:** A large-scale construction project in a geographically challenging region utilized AI for risk management. The AI system analyzed historical construction data, weather patterns, geological conditions, and other relevant factors to identify potential risks. The system then provided risk assessments, prioritization of mitigation strategies, and even generated contingency plans for various scenarios.
- **Benefits:** This AI-powered risk management system helped the construction team proactively address potential risks, minimizing delays and cost overruns. By identifying and mitigating risks early on, the project stayed on track, delivering a successful outcome within budget and timelines.
- **Challenges:** Integrating AI for risk management requires access to comprehensive data, including historical project records, weather forecasts, and site-specific information. The AI system must be continuously updated and validated to ensure accurate risk assessments and effective mitigation strategies.

4. AI in Civil Infrastructure Maintenance and Inspection:

Maintaining and inspecting aging civil infrastructure is a crucial but often challenging task. Visual inspection techniques are labor-intensive, time-consuming, and sometimes unreliable. AI is now transforming this process, making it more efficient and accurate.

- **Case Study:** A city government implemented an AI-powered system for bridge inspection. The system utilized drones equipped with high-resolution cameras to capture detailed images of bridge structures. Advanced image recognition algorithms then analyzed the images, identifying potential cracks, corrosion, and other structural defects.
- **Benefits:** The AI-powered inspection system significantly reduced the time and cost of bridge inspections compared to traditional methods. It also improved the accuracy and reliability of defect detection, leading to more targeted maintenance and repair efforts, extending the lifespan of bridges and ensuring public safety.
- **Challenges:** The development and deployment of AI-powered inspection systems require expertise in computer vision, drone technology, and data analysis. Additionally, ensuring the accuracy and reliability of the AI system requires rigorous testing and validation using real-world data.

5. AI for Smart Construction Site Management:

Construction sites are dynamic environments where real-time data and intelligent decision-making are paramount. AI is making construction sites smarter and more efficient.

- **Case Study:** A construction company deployed AI-powered sensors and cameras throughout a large-scale construction site. The sensors collected data on worker location, equipment utilization, material inventory, and environmental conditions. The AI system used this data to optimize logistics, manage equipment allocation, track progress, and identify potential safety hazards.

- **Benefits:** The AI-powered site management system improved the efficiency of construction operations, reduced delays, and enhanced worker safety. It provided real-time insights into site activities, allowing managers to make informed decisions, allocate resources effectively, and optimize workflows.
- **Challenges:** Implementing an AI-powered site management system requires careful planning, coordination, and integration with existing infrastructure. Data security and privacy concerns must be addressed, and ensuring the system's accuracy and reliability requires rigorous testing and validation.

Challenges and Considerations

While the benefits of AI in engineering are evident, it's crucial to acknowledge the challenges and considerations that accompany its implementation:

- **Data Requirements:** AI systems rely heavily on data. Acquiring, cleaning, and preparing large datasets for training AI algorithms can be time-consuming and resource-intensive.
- **Skill Gaps:** Successfully implementing and managing AI solutions in engineering projects requires skilled professionals with expertise in AI, data science, and

engineering disciplines. Bridging the skill gap is essential for maximizing the potential of AI.

- **Ethical Considerations:** AI systems, like any technology, are subject to ethical considerations. Addressing issues of bias, privacy, transparency, and responsibility in AI development and deployment is crucial for responsible and equitable use.
- **Integration Challenges:** Integrating AI solutions into existing engineering workflows and systems can be complex and require careful planning and execution to avoid disruptions and ensure compatibility.

Conclusion

The case studies presented showcase the real-world applications of AI in engineering projects, demonstrating its potential to transform design, construction, operation, and maintenance processes. While challenges exist, the benefits of increased efficiency, improved safety, reduced costs, and innovative solutions are undeniable. As AI continues to evolve, its impact on engineering project management will only grow, shaping the future of this critical industry.

The Future of AI in Engineering

The future of AI in engineering is full of exciting possibilities. We can expect to see further advancements in:

Generative design: AI algorithms will become even more sophisticated, capable of designing complex structures with unprecedented efficiency and innovation.

Simulation and modeling: AI-powered simulations will become increasingly realistic, allowing engineers to test and optimize designs in virtual environments.

Data-driven decision-making: AI will play an even greater role in providing data-driven insights and supporting informed decisions in engineering projects.

AI in IT Project Management

Project Planning and Resource Allocation

In the realm of IT project management, where intricate software development lifecycles and complex technological landscapes intertwine, the application of artificial intelligence (AI) has emerged as a transformative force. AI's ability to analyze vast amounts of data, identify patterns, and automate processes has revolutionized how project managers approach planning, resource allocation, and task prioritization.

Imagine a world where project timelines are no longer shrouded in uncertainty but are instead guided by data-driven insights. AI can analyze historical project data, taking into account factors like team performance, past project complexities, and external market conditions, to predict project timelines with remarkable accuracy. This empowers project managers to set realistic deadlines, allocate resources efficiently, and manage expectations effectively.

But AI's capabilities extend far beyond mere prediction. It can actively optimize resource allocation, taking into account individual skill sets, availability, and project requirements. By analyzing project tasks, AI can identify the best-suited team members for each assignment, ensuring optimal utilization of resources and maximizing efficiency.

Let's delve into a real-world scenario to illustrate this concept. Consider a large-scale software development project involving a diverse team of engineers, designers, and testers. Traditionally, resource allocation would involve manual estimations based on past experience and subjective judgments. However, AI-powered resource allocation tools can analyze project requirements, individual skill levels, and past performance data to generate an optimized team structure. This data-driven approach ensures that the right people are assigned to the right tasks, maximizing productivity and minimizing bottlenecks.

Furthermore, AI's ability to process vast datasets unlocks a new dimension of task prioritization. By analyzing project dependencies, deadlines, and team availability, AI can prioritize tasks based on their criticality and impact on the overall project. This dynamic prioritization ensures that high-impact tasks are addressed first, minimizing potential delays and ensuring project success.

Let's consider a hypothetical example of an IT project involving multiple software modules. AI can analyze the dependencies between these modules, identifying critical paths that need to be addressed first. This analysis might reveal that a particular module, while not necessarily the largest in scope, is crucial for the successful integration of other modules. By prioritizing this module, the AI-powered system ensures that dependencies are met and the overall project progresses efficiently.

The power of AI lies in its ability to analyze data, identify patterns, and learn from past experiences. This allows it to continuously refine its predictions, optimize resource allocation, and prioritize tasks based on evolving project needs. Imagine a project manager, armed with AI-powered insights, able to anticipate potential risks before they materialize, proactively adjust resource allocation, and make informed decisions to ensure project success.

Beyond these specific applications, AI has the potential to transform the very fabric of project management. It can facilitate seamless communication between team members, automate routine tasks, and provide project managers with real-time dashboards that offer comprehensive project overviews.

However, the integration of AI into project management isn't without its challenges. Data privacy concerns, the need for skilled professionals to interpret AI-generated insights, and the ethical implications of relying on AI for critical decisions are all issues that need to be carefully considered. Nevertheless, the potential benefits of AI in IT project management are undeniable. By embracing this transformative technology, project managers can unlock a new era of efficiency, productivity, and innovation.

Let's consider an example of a project involving the development of a mobile application for a major financial institution. With AI-powered tools, project managers can analyze user feedback from previous app releases, identify common pain points, and prioritize features that address these issues. AI can also automate code generation, ensuring consistency and reducing development time. By integrating AI into the development process, project managers can deliver high-quality, user-friendly applications in a shorter timeframe.

As AI continues to evolve, its role in IT project management will only grow more prominent. From predictive analytics and automation to decision support and collaboration, AI empowers project managers to navigate the complexities of IT projects with greater confidence and efficiency.

Imagine a future where project management is not just about managing tasks but about harnessing the power of AI to unlock a world of possibilities. This is the future that AI is ushering in for IT project management, a future where data-driven insights, automation, and intelligent decision-making converge to redefine project success.

Real World Examples

Case Study: Real-World Example of AI in IT Project Manangement

The realm of IT project management has witnessed a surge in AI adoption, transforming how teams approach development, security, and overall efficiency. Let's delve into real-world examples showcasing the impact of AI on various IT projects.

Project Planning and Resource Allocation: AI-Powered Predictive Analytics

AI-powered predictive analytics is transforming project planning and resource allocation in IT projects. By analyzing historical data, project dependencies, and team capacity, AI algorithms can forecast timelines, identify potential delays, and recommend optimal resource allocation strategies.

Impact

- **Improved project predictability:** AI-driven forecasting helps project managers better estimate timelines and anticipate potential risks, enhancing project predictability.
- **Optimized resource utilization:** By analyzing team skills, workload, and project demands, AI algorithms can allocate resources effectively, minimizing overutilization and underutilization.
- **Data-driven decision-making:** AI provides insights into project progress, resource utilization, and potential risks, enabling data-driven decision-making for project managers.

Challenges and Considerations

While AI offers remarkable benefits in IT project management, it's essential to acknowledge the challenges and considerations associated with its implementation:

- **Data Privacy and Security:** AI systems rely on vast amounts of data, raising concerns about data privacy and security. Organizations must ensure responsible data management practices to protect sensitive information.
- **Bias and Fairness:** AI algorithms can inherit biases present in the data they are trained on, potentially

leading to unfair or discriminatory outcomes. It's crucial to address these biases and ensure fairness in AI applications.

- **Transparency and Explainability:** AI decisions can be complex and opaque, making it difficult to understand how they are made. Ensuring transparency and explainability in AI models is essential for trust and accountability.
- **Skilled Professionals:** Implementing and managing AI systems requires skilled professionals with expertise in AI, data science, and project management. Bridging the skills gap is crucial for successful AI adoption.

CONCLUSION

AI is revolutionizing IT project management, offering a range of benefits for development, security, and efficiency. As AI continues to evolve, its impact on IT project management will only grow, transforming the industry and shaping the future of software development and IT services.

CHAPTER 3

AI IN CONSTRUCTION PROJECT MANAGEMENT

CONSTRUCTION PROJECT MANAGEMENT

AI is also being leveraged to automate various aspects of construction project management. AI-powered algorithms can analyze vast amounts of data, identify potential risks and allocate resources effectively. These algorithms are helping project managers to make more informed decisions, improve project planning, and ensure the timely and successful completion of construction projects. AI is assisting with construction planning and scheduling, cost estimating and budgeting, predictive analysis, project risk management, and quality control.

CONSTRUCTION PLANNING AND SCHEDULING

Imagine a construction project, a complex symphony of moving parts orchestrated to deliver a breathtaking structure. But what happens when unexpected delays disrupt the harmony, causing costs to soar and deadlines to slip? This is where the transformative power of AI steps in, revolutionizing the way we plan and schedule construction projects.

AI-powered scheduling tools are not just about crunching numbers; they are about understanding the intricate dance of variables that influence a project's trajectory. These tools analyze vast datasets, encompassing historical project data, weather patterns, material availability, and even contractor performance. By weaving these insights into a sophisticated tapestry, AI can anticipate potential delays before they occur, empowering project managers to make proactive adjustments.

For example, AI algorithms can identify potential bottlenecks in the supply chain, alerting managers to procure materials well in advance. By analyzing weather patterns, AI can predict potential rain delays, allowing for adjustments to the schedule or even the deployment of weather-resistant technologies. And by assessing contractor performance data, AI can identify areas for improvement, ensuring that the right resources are allocated to optimize progress.

But the magic of AI doesn't stop at predicting delays. It also extends to resource allocation, ensuring that the right personnel and equipment are deployed at the right time. AI algorithms can analyze project requirements and contractor capabilities, suggesting the most efficient allocation of resources. This not only improves project efficiency but also optimizes costs, preventing overstaffing and ensuring that resources are utilized effectively.

Here's how AI is transforming construction scheduling in practice:

Dynamic Scheduling and Real-Time Updates: AI-powered scheduling tools can dynamically adjust schedules in response to unforeseen events, providing real-time updates to stakeholders. This ensures that everyone is on the same page, reducing confusion and fostering better collaboration.

Automated Task Assignment: AI can analyze project requirements, skill sets, and contractor availability to automatically assign tasks, streamlining workflow and optimizing resource utilization.

Predictive Maintenance: By analyzing equipment usage data and identifying potential wear and tear patterns, AI can predict maintenance needs before equipment failures occur, minimizing downtime and ensuring project continuity.

Resource Optimization: AI can analyze project requirements, resource availability, and historical data to optimize resource allocation, minimizing costs and maximizing efficiency.

The benefits of AI-driven construction planning and scheduling are undeniable. By predicting delays, optimizing resource allocation, and providing real-time updates, AI empowers project managers to make informed decisions, leading to:

Reduced Project Delays: AI's ability to identify and mitigate potential delays significantly reduces the risk of project delays, ensuring timely completion.

Improved Cost Management: By optimizing resource allocation and predicting potential cost overruns, AI helps maintain project budgets and minimize financial risks.

Enhanced Project Efficiency: AI-powered scheduling tools streamline workflows, automate tasks, and optimize resource utilization, increasing overall project efficiency and productivity.

Better Communication and Collaboration: Real-time updates and automated communication tools foster transparency and enhance collaboration between stakeholders.

The impact of AI on construction scheduling is not just about efficiency; it's about transforming the very essence of project management. It's about moving beyond reactive responses to proactively addressing challenges, ensuring project success in an increasingly dynamic and complex world.

Implementing AI in Construction Planning and Scheduling: Practical Steps

Integrating AI into construction planning and scheduling requires careful consideration and a strategic approach. Here are some practical steps to guide the implementation process:

1. **Identify Business Needs:** Clearly define your objectives for implementing AI in scheduling. What specific challenges are you looking to address? Are you aiming to reduce delays, optimize resource allocation, or enhance communication?
2. **Data Acquisition and Preparation:** AI thrives on data. Gather relevant data from previous projects, including timelines, costs, resource utilization, and contractor performance. Ensure that your data is clean, accurate, and consistent.
3. **Choose the Right AI Tools:** Select AI-powered scheduling tools that align with your business needs and data availability. Consider factors like ease of use,

integration capabilities, and the platform's ability to scale with your project requirements.

4. **Pilot Projects and Gradual Implementation:** Start with pilot projects to test the capabilities of AI tools before widespread adoption. This allows you to identify potential challenges, refine processes, and gain valuable experience.

5. **Training and Education:** Provide training to your team on using AI tools effectively. This includes understanding data input requirements, interpreting AI-generated insights, and utilizing the platform's features.

6. **Continuous Monitoring and Improvement:** Constantly monitor the performance of AI tools and identify areas for improvement. Adjust your approach based on insights gained, ensuring that AI effectively meets your project management objectives.

By embracing AI, construction professionals can navigate the complexities of project scheduling with greater confidence, ensuring timely completion, optimized resource utilization, and enhanced project outcomes. AI is not simply a tool; it's a strategic partner, empowering project managers to orchestrate the construction symphony with greater precision, efficiency, and success.

Cost Estimation and Budgeting

The realm of construction project management is notorious for its intricate complexities, with cost estimation and budgeting being particularly sensitive areas. Traditional methods often rely on historical data analysis, expert judgment, and guesswork, leading to potential inaccuracies and cost overruns. This is where the transformative power of AI shines. AI algorithms, fueled by vast datasets and sophisticated analytical capabilities, can revolutionize the way we approach cost estimation and budgeting in construction projects.

Imagine a scenario where you're embarking on a new skyscraper project. Traditional methods might involve sifting through countless spreadsheets, historical data, and expert opinions to arrive at an initial cost estimate. But with AI, the process becomes significantly streamlined. AI-powered systems can analyze historical data from past projects, factoring in variables like material costs, labor rates, location, and even weather conditions, providing a much more accurate and reliable cost estimate.

Here's how AI excels in cost estimation and budgeting:

Data-Driven Accuracy: AI algorithms can analyze vast amounts of historical data from previous projects, including cost breakdowns, market trends, and economic indicators. This comprehensive data analysis enables AI to generate more accurate cost estimates than traditional methods, minimizing the risk of overestimation or underestimation.

Risk Identification and Mitigation: By analyzing data patterns and trends, AI can identify potential cost overruns, including those arising from unforeseen delays, material shortages, or labor disputes. This early detection allows project managers to proactively develop mitigation strategies, reducing the likelihood of budget blowouts.

Real-Time Cost Tracking and Forecasting: AI-powered systems can continuously track project costs in real time, comparing actual expenditures against the initial budget. This enables project managers to identify potential cost overruns early on and adjust their budgets accordingly, minimizing the impact of unforeseen circumstances.

Dynamic Budgeting and Resource Allocation: AI can analyze project data to optimize resource allocation, ensuring that resources are deployed effectively and efficiently. AI algorithms can identify areas where costs can be reduced without compromising project quality, leading to more optimized budgets and enhanced profitability.

Scenario Planning and What-If Analysis: AI can perform scenario planning, allowing project managers to assess the potential impact of different variables on project costs. This "what-if" analysis can help in making informed decisions regarding resource allocation, material selection, and contingency planning, minimizing the risk of cost overruns.

Let's delve into real-world examples to illustrate how AI is transforming cost estimation and budgeting in construction:

Automated Material Cost Estimation: AI-powered systems can analyze real-time market data, including material prices, supply chain dynamics, and weather conditions, to provide highly accurate material cost estimates. This eliminates the need for manual research and reduces the risk of inaccurate cost projections.

Labor Cost Optimization: AI can analyze labor availability, skill sets, and historical labor rates to provide optimized labor cost estimates. AI can also identify potential labor shortages and suggest proactive measures to mitigate their impact on project costs.

Predictive Modeling for Cost Overruns: AI can analyze historical project data to develop predictive models that identify potential cost overruns based on specific project parameters. This allows project managers to proactively take preventive measures, such as adjusting schedules, revising resource allocation, or negotiating better contracts.

AI-Powered Budget Management Platforms: Several software platforms are emerging that leverage AI to streamline budget management, automate cost tracking, and provide real-time insights. These platforms empower project managers with data-driven tools to make informed decisions and effectively manage project budgets.

Beyond these specific examples, AI is playing a pivotal role in transforming the entire construction industry, from the initial planning stages to the final handover. By automating repetitive tasks, providing accurate cost estimates, and offering valuable insights into project risks, AI is empowering project managers to make informed decisions, optimize budgets, and ultimately deliver projects on time and within budget.

As AI continues to evolve, we can expect even more innovative applications in cost estimation and budgeting. AI-powered systems will become more sophisticated, analyzing even more complex data and offering more nuanced insights to project managers. This will further enhance the efficiency, accuracy, and reliability of cost estimations, leading to better project outcomes and greater profitability for construction projects.

PROJECT RISK MANAGEMENT

Project risk management is an essential aspect of any project, especially in construction where complex variables and unpredictable factors can significantly impact project outcomes. Traditional risk management methods often rely on expert opinions and historical data, which can be subjective and prone to human error. However, AI-powered risk management offers a more data-driven and sophisticated approach, enabling project managers to identify potential risks, assess their impact, and develop effective mitigation strategies.

AI'S ROLE IN RISK IDENTIFICATION

One of the most significant contributions of AI to risk management is its ability to identify potential risks that might otherwise go unnoticed. AI algorithms can analyze vast amounts of data, including historical project data, weather patterns, market trends, and even social media sentiment, to identify potential risk factors that could impact the project's success.

Data Analysis for Pattern Recognition: AI algorithms excel at finding patterns and anomalies in data that might not be readily apparent to human analysts. By analyzing historical project data, AI can identify recurring risks, such as delays caused by specific weather conditions or material shortages. This data-driven approach allows project managers to anticipate and plan for these risks proactively.

Predictive Modeling for Early Risk Detection: AI-powered predictive models can analyze current project conditions and external factors to forecast potential risks. For example, an AI model could analyze real-time weather data, construction progress, and supplier performance to predict potential delays or supply chain disruptions. This proactive approach enables project managers to take corrective actions before these risks materialize.

Predictive analytics is a key area where AI is making a significant impact on construction project management. By analyzing historical data, project managers can use AI to predict potential delays, identify potential cost overruns, and anticipate other project challenges. These insights allow project managers to take proactive measures to mitigate risks and ensure the project stays on track.

Natural Language Processing for Identifying Risks in Textual Data: AI's natural language processing capabilities allow it to analyze textual data, such as project documents, emails, and online discussions. This analysis helps identify potential risks hidden within project plans, contracts, and stakeholder communications.

Assessing Risk Impact

Once potential risks are identified, AI can help project managers assess their impact on the project. This assessment involves determining the likelihood of the risk occurring and its potential consequences.

Quantitative Risk Assessment: AI algorithms can use historical data and statistical modeling to quantify the likelihood and impact of identified risks. For instance, an AI model could analyze historical data on construction delays caused by specific weather events to estimate the probability and severity of similar delays in the future.

Qualitative Risk Assessment: While quantitative assessment is valuable, AI can also assist in qualitative risk assessment. By analyzing project documentation, stakeholder feedback, and industry trends, AI can provide insights into the potential consequences of a risk that may not be quantifiable.

DEVELOPING MITIGATION STRATEGIES

AI's ability to analyze data and identify patterns enables it to develop effective risk mitigation strategies.

Generating Mitigation Options: Based on identified risks and their potential impact, AI can recommend various mitigation strategies. These options could include altering project schedules, procuring alternative materials, securing additional resources, or adjusting project plans.

Evaluating Mitigation Effectiveness: AI can evaluate the effectiveness of proposed mitigation strategies by simulating various scenarios. This allows project managers to compare the potential outcomes of different strategies and select the most effective one.

CASE STUDY: AI IN CONSTRUCTION RISK MANAGEMENT

A prominent construction company in the United States used AI to mitigate risks associated with labor shortages and supply chain disruptions. By analyzing historical data, market trends, and industry news, the AI system identified a potential risk of increased labor costs and material delays due to a growing demand for construction workers. The AI system then generated a series of mitigation strategies, including implementing early procurement for critical materials, exploring alternative construction technologies that require fewer skilled laborers, and proactively engaging with local trade schools to build a pipeline of future workers. The company implemented these strategies, successfully mitigating the potential impact of labor shortages and supply chain disruptions.

Benefits of AI in Construction Risk Management

Improved Accuracy and Objectivity: AI-powered risk assessment eliminates human bias and subjectivity, providing more accurate and reliable risk estimates.

Proactive Risk Management: AI allows for early identification and proactive mitigation of risks, minimizing potential delays and cost overruns.

Enhanced Decision-Making: AI-generated insights provide data-driven support for informed risk management decisions.

Reduced Costs and Increased Efficiency: By identifying and mitigating risks early, AI helps minimize project costs and increase efficiency.

Continuous Learning and Improvement: AI systems learn from past projects and adapt their risk assessment and mitigation strategies based on new data, continuously improving their effectiveness.

Challenges and Considerations

Data Availability and Quality: AI models rely heavily on accurate and comprehensive data. The availability and quality of data can significantly impact the effectiveness of AI-powered risk management.

Transparency and Explainability: While AI can provide insightful recommendations, it is crucial to understand the rationale behind these recommendations. Transparency and explainability in AI models are essential to building trust and ensuring accountability.

Human Expertise Remains Critical: AI is a valuable tool but should not replace human expertise. Project managers should use AI as a powerful complement to their knowledge and experience, interpreting and applying AI insights effectively.

Conclusion

AI is transforming project risk management in the construction industry, offering powerful tools to identify, assess, and mitigate risks. By leveraging data analysis, predictive modeling, and other AI capabilities, project managers can proactively address potential challenges, reduce project risks, and achieve better outcomes. However, it is important to acknowledge the limitations and challenges of AI, ensuring that it is used responsibly and effectively alongside human expertise.

Quality Control

Quality control and inspection are critical aspects of any construction project, ensuring that the finished product meets the specified standards and adheres to safety regulations. Traditional methods often involve manual inspections, which can be time-consuming, prone to human error, and potentially unsafe. AI is transforming how quality control and inspection are conducted in construction, offering numerous benefits and enhancing the efficiency and reliability of this process. Quality control is another area where AI is making a significant contribution to the construction industry. AI-powered algorithms can analyze images and videos of construction sites to identify potential defects or inconsistencies. This helps to ensure that construction projects meet the required quality standards and minimize the need for costly rework.

AI-Powered Visual Inspection and Defect Detection

One of the most impactful applications of AI in construction quality control is visual inspection and defect detection. AI-powered systems can analyze images and videos captured on construction sites, identifying potential defects and anomalies that might be overlooked by human inspectors. These systems are trained on vast datasets of images, enabling them to learn patterns and identify deviations from expected standards.

How It Works:

- **Image and Video Analysis:** AI algorithms process images and videos from construction sites, analyzing features like cracks, gaps, uneven surfaces, and misaligned components.
- **Defect Detection:** The algorithms identify potential defects by comparing the analyzed images and videos against established quality standards and detecting deviations.
- **Real-Time Reporting and Alerting:** The AI system generates reports and alerts project managers

and inspectors in real-time, highlighting areas requiring attention and facilitating prompt corrective action.

Examples:

- **Drones with AI-Powered Cameras:** Drones equipped with high-resolution cameras can capture aerial images of construction sites, providing a comprehensive view of the entire structure. AI algorithms analyze these images to identify potential defects like cracks in concrete, misaligned steel beams, or missing roofing tiles.
- **AI-Enabled Smart Glasses:** Workers can wear smart glasses equipped with AI-powered image recognition software. These glasses analyze the surrounding environment in real-time, alerting workers to potential hazards or defects, such as loose wires or improper installation.

Benefits of AI-Powered Visual Inspection:

- **Increased Efficiency:** AI-powered systems can process images and videos much faster than human inspectors, reducing the time required for inspections.
- **Improved Accuracy:** AI algorithms are less prone to human error and can identify subtle defects that might be missed by manual inspection.
- **Enhanced Safety:** By detecting potential hazards and defects early, AI systems can improve safety on construction sites, preventing accidents and injuries.
- **Real-Time Insights:** AI provides real-time feedback

and alerts, enabling faster corrective actions and preventing costly rework later in the project.

AI for Material Testing and Quality Assurance

Beyond visual inspections, AI is playing a crucial role in material testing and quality assurance. AI algorithms can analyze data from material testing equipment, identifying potential issues with materials and ensuring that they meet the project's specifications.

How It Works:

- **Data Analysis:** AI algorithms analyze data from material testing equipment, such as tensile strength tests, compression tests, and chemical analysis results.
- **Quality Assessment:** The algorithms compare the test data against established quality standards and identify materials that deviate from specifications.
- **Predictive Modeling:** AI models can be trained to predict the performance of materials over time, allowing for proactive material selection and mitigating potential risks.

Examples:

- **Concrete Strength Analysis:** AI can analyze data from concrete strength tests, predicting the long-term performance of concrete and identifying potential issues like cracking or early deterioration.

- **Steel Fatigue Analysis:** AI algorithms can analyze data from steel fatigue tests, identifying potential failure points and predicting the remaining lifespan of steel structures.

Benefits of AI in Material Testing:

- **Faster and More Efficient Testing:** AI automates data analysis, reducing the time required for material testing and enabling faster decision-making.
- **Enhanced Accuracy and Reliability:** AI algorithms minimize human error, ensuring more accurate and reliable results in material testing.
- **Predictive Maintenance and Asset Management:** AI can predict the lifespan and performance of materials, facilitating proactive maintenance and asset management strategies.

AI for Compliance Monitoring and Documentation

AI is also transforming how construction projects comply with regulations and standards. AI-powered systems can automate the process of monitoring compliance, generating reports, and documenting project progress.

How It Works:

- **Regulation and Standard Integration:** AI systems can be trained on relevant regulations and construction standards, enabling them to identify potential non-compliance issues.

- **Document Analysis:** AI algorithms can analyze construction documents, including blueprints, permits, and inspection reports, identifying inconsistencies and potential violations.
- **Automatic Reporting:** AI systems can automatically generate compliance reports, highlighting areas of concern and tracking progress towards meeting regulatory requirements.

Examples:

- **Permit Compliance Monitoring:** AI can analyze construction permits and compare them against project progress, ensuring that all necessary permits are in place and that construction activities align with permit requirements.
- **Safety Standard Compliance:** AI can analyze safety data from construction sites, identifying potential violations of safety standards and recommending corrective actions.

Benefits of AI in Compliance Monitoring:

- **Automated Compliance Tracking:** AI simplifies the process of monitoring compliance, reducing manual effort and ensuring accurate tracking of project progress.
- **Real-Time Alerting and Reporting:** AI systems provide real-time alerts when potential non-compliance issues are detected, facilitating prompt corrective actions.

- **Reduced Risk of Legal Issues:** By ensuring project compliance with regulations, AI helps reduce the risk of fines, legal disputes, and project delays.

Overcoming Challenges and Ethical Considerations

While AI offers significant benefits in quality control and inspection, there are challenges and ethical considerations to address:

- **Data Privacy and Security:** Ensuring the security and privacy of sensitive construction data collected by AI systems is crucial.
- **Bias and Fairness:** Training AI models on biased data can lead to biased outputs. Ensuring fairness and preventing biased decisions in quality control requires careful consideration.
- **Human Oversight and Validation:** AI systems should not replace human oversight entirely. It's essential to have experienced professionals validate AI-generated insights and make informed decisions.

The Future of AI in Construction Quality Control

The future of AI in construction quality control looks promising. Continued advancements in AI technologies will enable even more sophisticated applications, including:

- **Predictive Maintenance and Failure Detection:** AI will become increasingly adept at predicting equipment failures and material deterioration, enabling proactive maintenance and risk mitigation.
- **Autonomous Inspection Robots:** The use of autonomous robots for inspections will increase,

providing a more efficient and cost-effective way to conduct inspections in hazardous or inaccessible areas.
- **AI-Powered Quality Assurance Systems:** AI will play a more integrated role in quality assurance systems, providing real-time feedback and alerts to guide construction processes and ensure project success.

As AI technologies continue to evolve, they will play a transformative role in improving the efficiency, accuracy, and safety of quality control and inspection in the construction industry. Embracing AI solutions will enable construction projects to meet higher quality standards, reduce risks, and achieve optimal outcomes.

Real World Examples of AI in Construction Projects

The construction industry is undergoing a digital transformation, and AI is at the forefront of this revolution. From optimizing project schedules to enhancing safety protocols, AI is proving to be a valuable asset in construction project management.

Case Study 1: Smart Construction Scheduling with AI

One of the most significant challenges in construction is managing complex schedules, especially in large-scale projects. Delays are common due to factors like weather, material shortages, and unforeseen site conditions. AI can address this challenge by providing intelligent scheduling solutions.

A leading construction company, Skanska, employed an AI-powered scheduling tool to optimize the construction of a major infrastructure project. The tool analyzed historical data from previous projects, weather patterns, and material availability to create a dynamic schedule that adjusted automatically to changing conditions. This resulted in:

- **Reduced Delays:** The AI-powered scheduler successfully predicted potential delays and proactively adjusted the schedule, minimizing disruptions and keeping the project on track.
- **Improved Resource Allocation:** By analyzing resource availability and project demands, the AI tool optimized resource allocation, ensuring that the right resources were available at the right time.
- **Enhanced Communication:** The AI-enabled scheduling platform provided real-time updates to all stakeholders, ensuring everyone was informed about schedule changes and potential delays.

Case Study 2: Predictive Maintenance with AI for Construction Equipment

Downtime of construction equipment can significantly impact project timelines and budgets. AI can play a crucial role in predictive maintenance, preventing equipment failures and minimizing downtime.

A major construction equipment manufacturer, Caterpillar, implemented an AI-powered predictive maintenance system for its construction equipment. The system analyzed sensor data from machines in real time, detecting patterns that indicated potential failures. This enabled:

- **Early Failure Detection:** The system proactively identified potential equipment failures before they occurred, allowing maintenance crews to address issues before they caused major problems.
- **Optimized Maintenance Schedules:** Based on the AI-generated insights, the company optimized maintenance schedules, preventing unnecessary downtime and extending the lifespan of its equipment.
- **Reduced Maintenance Costs:** By preventing major equipment breakdowns, the system significantly reduced repair costs and minimized unexpected expenses.

Case Study 3: AI-Powered Quality Control and Inspection

Maintaining quality standards is essential in construction projects. AI can automate quality control processes, ensuring that projects meet specifications and minimize defects.

A renowned construction company, Bechtel, utilized an AI-powered system for quality control and inspection on a major infrastructure project. The system analyzed images and videos from construction sites, detecting defects and anomalies in real time. This resulted in:

- **Automated Defect Detection:** The AI system accurately detected defects in concrete, welding, and other construction elements, reducing the need for manual inspections.
- **Increased Inspection Efficiency:** The automated inspection process significantly sped up the inspection process, allowing quality control teams to focus on more complex issues.

- **Improved Project Quality:** By identifying and addressing defects early, the AI system ensured that the project met the highest quality standards.

Case Study 4: AI-Driven Construction Safety

Safety is paramount in construction, and AI can enhance safety protocols, identifying potential hazards and promoting safer work environments.

A leading construction company, Turner Construction, adopted an AI-powered system to monitor worker safety on a large-scale construction project. The system analyzed video footage from the site, detecting unsafe practices, such as workers not wearing safety gear or operating machinery improperly. This resulted in:

- **Proactive Hazard Identification:** The AI system proactively identified potential hazards, alerting safety officers to intervene and mitigate risks.
- **Improved Worker Safety:** By detecting unsafe practices, the system helped prevent accidents and injuries, creating a safer work environment for construction workers.
- **Data-Driven Safety Improvements:** The system collected data on safety incidents, allowing the company to analyze trends and implement data-driven safety improvements.

Case Study 5: AI in Cost Estimation and Budgeting

Accurate cost estimations are crucial for successful construction projects. AI can analyze historical data, market trends, and project specifications to provide more accurate cost esti-mations.

A global construction company, AECOM, utilized an AI-powered cost estimation tool to improve budget accuracy on a major infrastructure project. The tool analyzed historical project data, market pricing trends, and project specifications to generate realistic cost estimates. This enabled:

- **Improved Budget Accuracy:** The AI-generated cost estimates were significantly more accurate than traditional methods, reducing the risk of budget overruns.
- **Enhanced Bid Management:** The tool provided insights into bidding strategies, helping the company submit competitive bids and win contracts.
- **Data-Driven Budgeting:** The tool collected data on project costs, allowing the company to analyze historical data and improve future budget forecasting.

These case studies demonstrate the transformative potential of AI in construction project management. By leveraging AI-powered tools and solutions, construction companies can opti-mize project schedules, enhance safety protocols, improve quality control, and achieve greater project success. As AI continues to evolve, its impact on the construction industry will only grow stronger.

However, it's important to acknowledge that AI implementa-tion is not without its challenges. Despite any challenges, the potential benefits of AI in construction are undeniable. To harness its power effectively, construction companies need to:

- **Invest in Data Infrastructure:** Companies should prioritize building robust data management systems that collect and organize high-quality data.
- **Develop AI Expertise:** They should invest in training and development programs to build a skilled workforce capable of managing and leveraging AI solutions.
- **Collaborate with AI Experts:** Partnering with AI experts and technology providers can accelerate AI adoption and address implementation challenges.
- **Focus on Ethical AI Development:** Ensure that AI solutions are developed and used responsibly, addressing data privacy, bias, and other ethical concerns.

By overcoming these challenges and embracing the potential of AI, the construction industry can achieve a new level of efficiency, safety, and sustainability. AI is not just a technological advancement but a transformative force that will shape the future of construction.

Challenges

The use of AI in construction is not without its challenges. One major concern is data privacy and the security of sensitive information collected by AI systems. It is essential that construction companies implement robust data security measures to protect the privacy of their workers and stakeholders.

Another challenge is the need for skilled professionals who can manage and interpret AI-powered systems. Construction companies will need to invest in training and development programs to ensure their workforce has the necessary skills to effectively utilize AI in their work.

Ethical considerations also play a crucial role in the use of AI in construction. It is essential to ensure that AI systems are developed and deployed in a way that is fair, unbiased, and does not discriminate against individuals or groups.

Despite these challenges, the potential benefits of AI in construction are undeniable. AI has the power to revolutionize the industry, making it more efficient, safe, and sustainable. By embracing AI technologies, construction companies can enhance project outcomes, reduce costs, and improve worker safety. The future of construction is AI-powered, and those who are willing to adapt and innovate will be the leaders of this transformative era.

CHAPTER 4

AI-POWERED TOOLS AND PLATFORMS

PROJECT MANAGEMENT SOFTWARE AND AI INTEGRATION

Project management software platforms are undergoing a transformation as AI integration becomes increasingly prevalent. This integration is revolutionizing how project managers approach tasks, enabling them to make better-informed decisions, improve efficiency, and enhance collaboration.

AI-POWERED PROJECT MANAGEMENT SOFTWARE

Several popular project management software platforms have embraced AI to augment their capabilities. These platforms are designed to streamline workflows, automate tasks, and provide valuable insights that empower project managers. Some of the key features commonly found in AI-integrated project management software include:

- **Predictive Analytics:** AI algorithms can analyze historical project data, such as timelines, budgets, and resource allocation, to predict future outcomes. This helps project managers identify potential delays, cost overruns, and resource bottlenecks early on, allowing them to proactively adjust plans and mitigate risks. For example, a project manager can leverage AI to predict the likelihood of a project being completed on time based on past performance data.

- **Automated Task Management:** AI can automate routine tasks, freeing up project managers to focus on strategic initiatives. This includes tasks such as scheduling meetings, sending reminders, assigning tasks, and generating reports. By automating these tasks, project managers can save time and reduce the risk of human error. For instance, an AI-powered assistant can automatically allocate tasks to team members based on their skills and availability, ensuring efficient workload distribution.

- **Resource Optimization:** AI can analyze project requirements and resource availability to optimize resource allocation. This can involve identifying the most suitable team members for specific tasks, predicting resource needs based on project milestones, and adjusting resource allocation dynamically to ensure project success. For example, an AI-powered tool can analyze the skills of team members and automatically assign them to tasks that match their expertise, maximizing efficiency.

- **Risk Management:** AI can assist in identifying and assessing project risks, helping project managers prioritize mitigation efforts. By analyzing historical data, project plans, and external factors, AI algorithms can predict potential risks and recommend strategies to minimize their impact. For example, an AI-powered risk management tool can identify potential delays based on weather patterns or supply chain disruptions, allowing project managers to implement contingency plans in advance.

- **Collaboration Enhancement:** AI can enhance communication and collaboration among project stakeholders. This includes features such as automated communication channels, real-time task updates, and AI-powered document sharing. For example, an AI-powered platform can summarize meeting discussions and generate action items, ensuring that everyone is aligned and informed.

Popular Platforms with AI Integration

- **Microsoft Project:** Microsoft Project, a well-known project management software, has integrated AI features, including predictive analytics, automated task management, and resource optimization. Its AI-powered features help streamline project workflows, improve efficiency, and provide data-driven insights.
- **Asana:** Asana, a cloud-based project management platform, offers AI-powered features to enhance task management, resource allocation, and collaboration. Its AI algorithms analyze project data to recommend improvements and automate repetitive tasks, freeing up project managers for strategic decision-making.

- **Monday.com:** Monday.com, a customizable project management platform, leverages AI to automate workflows, provide intelligent insights, and enhance collaboration. Its AI-powered features include automated task assignments, intelligent resource allocation, and predictive analytics.
- **Jira:** Jira, a widely used project management tool for software development, incorporates AI features for workflow automation, issue tracking, and code analysis. AI algorithms analyze project data to identify patterns, recommend solutions, and improve team efficiency.
- **Smartsheet:** Smartsheet, a cloud-based work management platform, integrates AI capabilities to automate workflows, optimize resource allocation, and provide insights for better decision-making. Its AI features include intelligent task assignment, predictive analytics, and automated reporting.

Case Studies of AI in Project Management Software

- **Construction Project Scheduling:** A construction company used AI-powered scheduling software to optimize project timelines. By analyzing historical data and weather patterns, the software predicted potential delays and recommended adjustments to the schedule. This enabled the company to complete projects on time and within budget.
- **IT Project Risk Management:** An IT firm implemented AI-powered risk management software

to identify and assess potential risks in software development projects. The software analyzed historical data, code quality metrics, and external factors to identify potential vulnerabilities. This helped the firm prioritize mitigation efforts and reduce the likelihood of project failures.

- **Engineering Project Cost Estimation:** An engineering firm used AI-powered cost estimation software to improve the accuracy of project budgets. The software analyzed historical data, market trends, and project specifications to generate more accurate cost estimations, reducing the risk of budget overruns.

Beyond Basic Features

AI integration in project management software is constantly evolving, with platforms incorporating more advanced features like:

- **Natural Language Processing (NLP):** NLP enables users to interact with project management software using natural language, making it easier to manage tasks, search for information, and receive insights. For example, a user could simply ask a virtual assistant, "What are the key risks for this project?" instead of manually sifting through data.
- **Machine Learning (ML):** ML algorithms can analyze large datasets to identify patterns, make predictions, and learn from new data. This allows project management software to provide more accurate insights, optimize workflows, and adapt to changing conditions. For example, ML can be used to predict the

likelihood of a project being completed on time, based on factors such as team performance, resource availability, and project complexity.

- **Computer Vision:** Computer vision enables project management software to analyze images and videos, identifying potential issues or opportunities. This can be used to automate quality inspections, monitor construction progress, and detect potential safety hazards. For example, computer vision can be used to detect defects in construction materials, improving quality control and reducing rework.
- **Chatbots:** Chatbots powered by AI can answer questions, provide support, and automate tasks, improving communication and efficiency. For example, a chatbot can be used to answer common questions about project status, deadlines, or task assignments, freeing up project managers to focus on other priorities.

Benefits of AI Integration

The integration of AI into project management software offers numerous benefits for project managers and their teams:

- **Increased Efficiency:** AI can automate tasks, streamline workflows, and eliminate manual errors, improving overall project efficiency.
- **Improved Accuracy:** AI algorithms can analyze data to provide more accurate insights and predictions, leading to better-informed decisions and fewer errors.
- **Enhanced Collaboration:** AI can enhance communication, task management, and collaboration

between project stakeholders, fostering a more productive and streamlined project environment.

- **Reduced Risk:** AI can help identify and assess risks early on, enabling project managers to take proactive steps to mitigate potential issues and reduce the likelihood of project failures.
- **Better Decision-Making:** AI provides data-driven insights, enabling project managers to make more informed decisions and optimize project outcomes.

CHALLENGES OF AI INTEGRATION

While AI integration offers significant benefits, it also presents several challenges:

- **Data Quality:** AI models are only as good as the data they are trained on. Ensuring the quality and accuracy of data is crucial for reliable insights.
- **Data Privacy:** Using AI involves collecting and analyzing data, raising concerns about data privacy and security. Organizations must implement robust data security measures to protect sensitive information.
- **Ethical Considerations:** AI algorithms can perpetuate biases present in the data they are trained on. Addressing ethical considerations and ensuring fairness in AI implementation is essential.
- **Skilled Professionals:** Successfully implementing AI in project management requires skilled professionals with expertise in AI, data science, and project management.

Overcoming the Challenges

To overcome the challenges of AI integration, organizations should:

- **Prioritize Data Quality:** Implement processes to ensure data accuracy, completeness, and consistency.
- **Address Data Privacy Concerns:** Implement robust data security measures and comply with relevant privacy regulations.
- **Foster Ethical AI Development:** Establish clear ethical guidelines for AI use and ensure transparency and accountability.
- **Invest in Training and Development:** Provide training for project managers and other stakeholders to equip them with the skills necessary to manage AI-driven projects.

Conclusion

The integration of AI into project management software is transforming how project managers work. By leveraging AI to automate tasks, gain insights, and enhance collaboration, project managers can improve efficiency, reduce risks, and achieve better outcomes. However, organizations need to address challenges such as data quality, privacy, and ethical considerations to ensure responsible and effective AI implementation. By embracing AI's potential and overcoming these challenges, organizations can unlock the transformative power of this technology to drive success in project management and beyond.

AI Enabled Data Analytics and Visualization Tools

The world of data is no longer just a sea of numbers; it's a rich tapestry woven with insights, patterns, and potential. AI-enabled data analytics and visualization tools have emerged as potent instruments for project managers, allowing them to decipher these intricate patterns and transform raw data into actionable intelligence. These tools are like a powerful magnifying glass, revealing hidden trends, potential risks, and opportunities that could otherwise slip through the cracks.

Imagine a construction project manager struggling to make sense of a massive dataset containing information about material costs, labor hours, weather conditions, and equipment availability. This information, if left raw, is a chaotic jumble. But with AI-powered data analytics, this manager can effortlessly unravel the complex web of data. The AI tool could identify correlations between weather delays and specific types of work, predict potential cost overruns based on material price fluctuations, or even suggest optimal resource allocation based on historical data. This level of insight empowers the manager to make proactive decisions, minimize risks, and ultimately achieve better project outcomes.

The power of AI-enabled data analysis goes beyond simple identification; it extends to visualization. These tools translate complex data into compelling visuals, allowing project managers to grasp complex trends at a glance. Imagine a dashboard displaying the progress of multiple projects in real-time, highlighting critical milestones, potential bottlenecks, and resource utilization. This visual representation empowers project managers to quickly identify areas needing attention, make informed adjustments, and ensure projects stay on track.

Here are some key features of AI-enabled data analytics and visualization tools:

- **Automated Data Cleaning and Preparation:** AI algorithms can automatically clean and prepare data, removing inconsistencies and errors, saving valuable time and effort for project managers.
- **Predictive Analytics:** By analyzing historical data and current trends, these tools can provide accurate forecasts of project timelines, potential delays, and resource requirements, enabling proactive risk management.
- **Pattern Recognition:** AI excels at identifying complex patterns within data, revealing hidden correlations and insights that would be difficult for humans to discern.
- **Real-Time Data Visualization:** Data visualization capabilities offer dynamic and interactive dashboards, providing real-time insights into project progress, resource utilization, and key performance indicators.
- **Actionable Insights:** AI-powered tools go beyond just analyzing data; they deliver actionable insights, identifying areas for improvement, recommending solutions, and guiding decision-making.

Let's explore some real-world examples of how these tools are transforming project management:

- **Construction Project Management:** In the construction industry, AI-powered data analytics tools can analyze historical data to predict material costs, identify potential supply chain disruptions, and optimize project schedules. For example, a tool might analyze weather data to identify patterns that affect specific construction tasks, leading to better scheduling and risk mitigation.

- **IT Project Management:** AI can analyze code repositories to identify potential bugs and vulnerabilities, predict software development timelines, and automate code quality checks. Imagine a tool that analyzes past project data to predict the likelihood of successful software releases, helping project managers make data-driven decisions regarding resource allocation and release schedules.

- **Engineering Project Management:** AI can analyze engineering designs to identify potential structural weaknesses, optimize material usage, and simulate different scenarios to evaluate project risks. For example, a tool could simulate the performance of a bridge design under different weather conditions, identifying potential weaknesses and suggesting design modifications.

The use of AI-enabled data analytics and visualization tools extends far beyond these examples. Their applications are becoming increasingly diverse, touching every aspect of project management. From predicting project budgets to optimizing resource allocation, these tools are proving to be indispensable assets for project managers seeking to improve efficiency, reduce risks, and deliver successful outcomes.

The power of AI in data analytics and visualization is not without its challenges. Project managers must consider the ethical implications of using AI, such as data privacy, bias, and responsible AI development. They must also ensure that they have the necessary skills and expertise to effectively manage and interpret AI-generated insights.

In conclusion, AI-enabled data analytics and visualization tools are revolutionizing project management, empowering managers with unprecedented insights and data-driven decision-making capabilities. By harnessing the power of AI, project managers can navigate the complexities of modern projects with greater confidence, optimize resource allocation, mitigate risks, and achieve groundbreaking results. As AI technology continues to evolve, we can expect even more transformative applications in the future, pushing the boundaries of project management and ushering in a new era of efficiency, innovation, and success.

AI POWERED COMMUNICATION AND COLLABORATION PLATFORMS

The realm of project management is undergoing a profound transformation, driven by the advent of artificial intelligence (AI). This revolutionary technology is poised to redefine how projects are planned, executed, and delivered, ushering in a new era of efficiency, collaboration, and innovation. AI is not just about replacing human roles; it is about empowering project managers, engineers, IT professionals, and construction specialists with tools that amplify their abilities and unlock new possibilities.

One of the most impactful areas where AI is making its mark is in communication and collaboration. Traditional project management methods often struggled with information silos, fragmented communication channels, and time-consuming coordination efforts. AI-powered platforms are emerging to bridge these gaps, fostering seamless communication and collaboration between all project stakeholders. These platforms are built on the principles of natural language processing (NLP) and machine learning (ML), enabling them to understand, interpret, and respond to human language in a way that facilitates efficient information exchange.

AI-Powered Communication Tools: Revolutionizing Project Conversations

Imagine a scenario where project updates, task assignments, and critical information are shared instantly and seamlessly across the entire project team. This is the power of AI-powered communication tools. These tools leverage NLP and ML to automate communication tasks, streamline information flow, and enhance the overall communication experience. Here are some key features of these tools:

- **Chatbots for 24/7 Assistance**: AI-powered chatbots are becoming indispensable virtual assistants for project teams. These chatbots can answer frequently asked questions, provide quick access to project documentation, and even handle basic task assignments, freeing up project managers' time for strategic initiatives.

- **Virtual Assistants for Task Management:** AI-powered virtual assistants can help project managers stay organized and on top of their tasks. These assistants can automatically schedule meetings, send reminders, and even prioritize tasks based on urgency and importance.

- **Automated Meeting Scheduling:** Scheduling meetings can be a time-consuming process, especially when coordinating schedules across multiple time zones and busy calendars. AI-powered scheduling tools analyze team members' availability and automatically suggest meeting times that work for everyone, saving precious time and effort.

- **Real-Time Translation for Global Collaboration:** In today's globalized world, projects often involve teams from diverse countries and cultures. AI-powered translation tools break down language barriers, facilitating seamless communication and collaboration between international teams.

- **Sentiment Analysis for Early Warning Signals:** AI can analyze communication patterns and identify sentiment shifts within a project team. This sentiment analysis can alert project managers to potential issues, conflicts, or areas of concern, allowing for proactive intervention and early resolution.

Enhancing Collaboration with AI-Powered Platforms

Beyond individual communication tools, AI is also transforming the way entire project teams collaborate. AI-powered platforms provide a centralized hub for communication, information sharing, and task management, enabling seamless coordination and collaboration across all project phases. Here are some key aspects of these platforms:

- **Shared Workspaces for Team Collaboration:** AI-powered platforms create virtual workspaces where project teams can collaborate on documents, share files, and communicate in real-time. These platforms provide a central repository for all project-related information, eliminating information silos and promoting transparency.

- **Task Management and Progress Tracking:** These platforms streamline task management, allowing team members to assign tasks, track progress, and update statuses in real-time. AI algorithms can automatically prioritize tasks based on deadlines and dependencies, ensuring efficient task completion.

- **Real-Time Communication Channels:** AI-powered platforms integrate various communication channels, such as instant messaging, video conferencing, and email, into a single interface. This centralized communication hub ensures that all team members are connected and can easily access relevant information and updates.

- **Knowledge Management and Information Sharing:** AI can analyze project documentation, meeting transcripts, and communication logs to extract key insights and create a knowledge base that can be accessed by the entire team. This knowledge management system ensures that everyone has access to relevant information and avoids redundancy.

- **AI-Powered Project Management Tools:** These platforms integrate AI-powered tools for project planning, risk assessment, budget forecasting, and resource allocation. By leveraging AI algorithms, these tools can provide data-driven insights, optimize decision-making, and enhance project outcomes.

Real-World Examples of AI in Collaboration and Communication

The transformative potential of AI-powered communication and collaboration platforms is already being realized in numerous industries. Here are some real-world examples:

- **Construction Projects:** AI-powered platforms are helping construction companies streamline communication between project teams, subcontractors, and suppliers. These platforms facilitate real-time updates on progress, material deliveries, and potential issues, enabling better coordination and minimizing delays.
- **Engineering Projects:** AI-powered platforms are supporting collaboration between engineers, designers, and contractors on complex projects. These platforms facilitate real-time data sharing, enabling better design decisions, optimized resource allocation, and streamlined construction processes.

Challenges and Future of AI in Communication and Collaboration

While the benefits of AI-powered communication and collaboration platforms are undeniable, it's important to acknowledge potential challenges:

- **Data Privacy and Security:** As these platforms collect and analyze vast amounts of project data, ensuring data privacy and security is paramount. Robust security measures must be in place to protect sensitive information.
- **Integration with Existing Systems:** Integrating AI platforms with existing project management systems and workflows can be complex and require careful planning and implementation.
- **Ethical Considerations:** AI algorithms can sometimes exhibit biases, leading to potential fairness issues. It's crucial to develop AI systems that are fair, transparent, and unbiased.
- **Human-AI Collaboration:** While AI can automate tasks and provide insights, it's essential to maintain a human-centric approach. Project managers need to understand AI capabilities, interpret its outputs, and ultimately make informed decisions.

Despite these challenges, the future of AI in project management is bright. As AI technology continues to advance, we can expect even more sophisticated tools and platforms that will further enhance communication, collaboration, and project outcomes. The key is to embrace AI as a powerful ally, leveraging its capabilities to augment human intelligence and drive project success. By embracing AI, we can unlock a new era of project management, where communication is seamless, collaboration is effortless, and innovation is limitless.

Beyond the Hype: Responsible Integration of AI in Project Management

The emergence of advanced AI technologies is poised to redefine the landscape of project management, offering groundbreaking solutions that can revolutionize how projects are planned, executed, and delivered. While these emerging AI technologies hold incredible promise for project management, it's crucial to approach their implementation with a balanced perspective. AI is not a magic bullet but rather a powerful tool that can enhance project management practices when integrated responsibly.

- **Focus on Problem Solving:** Instead of simply adopting AI for the sake of it, identify specific project challenges that AI can effectively address.
- **Data Quality and Integrity:** AI relies on accurate and relevant data. Invest in data quality management to ensure that AI models are trained on reliable data, leading to more accurate predictions and insights.
- **Ethical Considerations:** Prioritize ethical considerations in AI development and deployment, addressing issues such as data privacy, bias, and responsible AI governance.
- **Human-AI Collaboration:** Recognize that AI is a tool to augment human capabilities, not replace them. Foster a collaborative environment where human expertise complements AI-driven insights.

Conclusion

Emerging AI technologies are poised to transform the world of project management, offering unprecedented opportunities to enhance efficiency, productivity, and innovation. By embracing these technologies with a strategic and responsible approach, project managers can leverage the power of AI to unlock new levels of success and drive positive change in their organizations.

Choosing the Right AI Tools for Your Project

The world of project management is rapidly evolving, fueled by the transformative power of artificial intelligence (AI). From streamlining tasks to making better decisions, AI is becoming an indispensable ally for project managers across industries. However, navigating the vast landscape of AI tools and platforms can be daunting. Following is a comprehensive guide to help you choose the right AI tools for your specific project needs, ensuring you make the most of this powerful technology.

Understanding Your Project Requirements

The first step in selecting the right AI tools is a thorough understanding of your project's specific requirements. Ask yourself:

What are your project goals? Are you aiming to improve efficiency, reduce costs, enhance quality, or achieve a specific outcome?

What are the key challenges you face? Are you struggling with resource allocation, risk management, or communication bottlenecks?

What types of data are you working with? Do you have access to structured data like project timelines, cost breakdowns, and resource availability? Or are you dealing with unstructured data such as emails, meeting transcripts, and project documents?

What level of AI expertise do you have? Are your team members familiar with AI concepts and tools? Or do you need user-friendly platforms with minimal technical expertise required?

Industry Best Practices and Considerations

Once you've defined your project requirements, consider the following industry best practices and considerations:

- **Scalability**: Choose tools that can scale with your project's growth and evolving needs. Avoid tools that are limited by capacity or functionality for larger projects.
- **Integration:** Ensure the AI tools you choose integrate seamlessly with your existing project management software and workflows. Look for platforms that offer APIs or connectors for easy integration.
- **Data Security and Privacy**: Prioritize tools that prioritize data security and privacy. Ensure compliance with relevant regulations and industry standards.
- **Transparency and Explainability**: Select AI tools that provide transparency into their decision-making processes. Look for tools that offer explanations for their predictions or recommendations.
- **User Friendliness**: Choose user-friendly interfaces that are intuitive and easy to navigate, even for team members with limited AI expertise.

TYPES OF AI TOOLS AND PLATFORMS

The AI landscape offers a diverse range of tools and platforms designed to support various project management aspects. Here's a breakdown of common categories:

1. Project Management Software with AI Integration:

Features: These platforms integrate AI capabilities directly into their project management functionalities, offering features like automated scheduling, risk assessment, resource allocation, and predictive analytics.

Examples:

- **Microsoft Project:** Leverages AI to optimize project schedules, predict delays, and recommend resources.
- **Asana:** Offers AI-powered features for task prioritization, workload management, and communication insights.
- **Smartsheet:** Incorporates AI for automated task assignments, resource allocation, and predictive forecasting.

2. AI-Enabled Data Analytics and Visualization Tools:

Features: These tools use AI for data analysis, visualization, and reporting, providing actionable insights for project management decisions.

Examples:

- **Power BI:** Offers AI-powered data discovery,

visualization, and forecasting capabilities for analyzing project data.

- **Tableau:** Provides AI-driven features for data exploration, visualization, and insights extraction from project datasets.
- **Google Analytics:** Uses AI to analyze website traffic and user behavior, providing insights for project marketing and communication strategies.

3. AI-Powered Communication and Collaboration Platforms:

Features: These platforms leverage AI to enhance team communication and collaboration, offering features like chatbots, virtual assistants, and automated meeting scheduling.

Examples:

- **Slack:** Integrates AI-powered bots for task management, project updates, and communication automation.
- **Microsoft Teams:** Offers AI features like automated meeting transcriptions, real-time translation, and intelligent search.
- **Zoom:** Incorporates AI for features like meeting transcription, automatic closed captioning, and background noise reduction.

4. Emerging AI Technologies for Project Management:

- **Features:** This category encompasses cutting-edge AI technologies that are transforming project management practices, such as:

- **Natural Language Processing (NLP):** Enables machines to understand and interpret human language, facilitating natural communication with AI systems.
- **Computer Vision:** Allows AI to "see" and analyze visual data, such as construction site images or engineering drawings.
- **Robotics and Automation:** Automates tasks using robots and AI-powered systems, improving efficiency and reducing human error.
- **Blockchain:** Provides a secure and transparent ledger for tracking project progress, payments, and material flow.
- **Internet of Things (IoT): Connects** devices and sensors to collect real-time data on project operations, enabling data-driven insights and proactive management.

Choosing the Right AI Tools for Your Project: A Step-by-Step Approach

1. **Define Your Project Objectives:** Clearly define your project goals, key challenges, and desired outcomes. This will guide your tool selection process.
2. **Assess Your Data:** Identify the types of data you have access to and determine if it's suitable for AI analysis. Consider data availability, quality, and format.
3. **Evaluate Tool Features:** Research different AI tools and platforms, focusing on features that align with your project requirements.
4. **Consider Integration Capabilities:** Ensure the

chosen tools integrate seamlessly with your existing project management software and workflows.

5. **Prioritize Data Security and Privacy:** Select tools that prioritize data security and privacy, ensuring compliance with regulations.

6. **Trial and Evaluation:** Take advantage of free trials or evaluation periods to test different tools and see how they perform in your specific project context.

7. **Gather Feedback:** Involve your team members in the selection process and gather their feedback on user experience and ease of use.

8. **Start Small and Scale Gradually:** Begin with a pilot project using the chosen AI tools. Gradually expand their use as you gain experience and confidence.

CONCLUSION

AI is a powerful tool for transforming project management practices, but choosing the right tools is crucial for success. By following the steps outlined above, you can select AI solutions that meet your specific project needs, empower your team, and achieve exceptional results. Embrace the power of AI to enhance your project management capabilities and propel your projects to new heights of efficiency, innovation, and success.

CHAPTER 5

AI FOR PREDICTIVE ANALYTICS IN PROJECT MANAGEMENT

FORECASTING PROJECT TIMELINES AND DEADLINES

Imagine a project manager staring at a complex spreadsheet, trying to predict when a critical milestone will be reached. The spreadsheet is filled with historical data, but it's overwhelming. The manager needs a way to sift through the data, identify patterns, and forecast the project timeline with greater accuracy. This is where artificial intelligence (AI) steps in, offering a transformative solution.

AI, particularly its ability to analyze vast amounts of data, can revolutionize the way we predict project timelines and identify potential delays. By feeding AI systems with historical project data, we can unlock powerful insights that traditional methods often miss. Let's explore how AI can help us achieve this:

1. **Unveiling the Power of Historical Data:**

77

Think of historical project data as a treasure trove of knowledge, waiting to be mined. It contains valuable insights about past projects, including:

- **Task durations:** How long did it take to complete specific tasks in previous projects? Did certain tasks consistently take longer than anticipated?
- **Resource allocation:** How were resources allocated in past projects? Were certain resources frequently overbooked or underutilized?
- **Risk factors:** What risks were encountered in past projects? How did these risks impact timelines and budgets?

AI can analyze this data and identify patterns that are difficult or impossible for humans to discern. For example, AI might reveal a correlation between certain types of tasks and their typical durations, or identify recurring risks that have historically caused delays.

2. Building Predictive Models with Machine Learning:

AI's core strength lies in its ability to learn from data and make predictions based on those learnings. Machine learning, a subset of AI, is particularly suited for creating predictive models for project timelines.

These models work by analyzing historical data and identifying relationships between different variables. For instance, a machine learning model might learn that:

- **Task complexity:** Projects with higher task complexity tend to take longer to complete.
- **Resource availability:** Delays are more likely when resources are scarce or overbooked.
- **External factors:** Weather conditions, supply chain disruptions, or regulatory changes can impact project timelines.

Once trained on historical data, these models can predict future project timelines with greater accuracy than traditional methods.

3. **Beyond Simple Predictions: Identifying Potential Delays:**

AI goes beyond simply predicting timelines; it can also help us identify potential delays before they occur. Here's how:

- **Early warning systems:** AI can monitor project progress in real-time and flag potential delays based on deviations from expected timelines or resource utilization patterns.
- **Risk analysis:** AI can analyze project data to identify potential risks that could impact timelines, such as supplier issues, weather events, or regulatory changes.
- **Scenario modeling:** AI can create different scenarios based on various risk factors and predict their potential impact on project timelines, allowing us to develop contingency plans.

4. **Examples of AI in Action:**

Let's look at some real-world examples of how AI is transforming project timeline prediction:

- **Construction projects:** AI-powered systems are being used to analyze weather data and predict potential delays due to rainfall or extreme temperatures. These insights help construction managers optimize schedules and allocate resources accordingly.
- **Software development:** AI can analyze historical data from software development projects to predict the time it takes to complete different stages, such as coding, testing, and deployment. This helps developers manage expectations and prioritize tasks effectively.
- **Engineering projects:** AI can analyze complex engineering data, such as simulations and test results, to identify potential design flaws or material weaknesses that could lead to delays or safety concerns.

5. **The Benefits of AI-Powered Timeline Predictions:**

The benefits of incorporating AI into project timeline prediction are undeniable:

- **Increased accuracy:** AI models can predict timelines with greater accuracy than traditional methods, reducing the risk of project delays and cost overruns.
- **Proactive risk management:** AI can identify potential delays before they occur, enabling us to take preventive measures and mitigate risks.
- **Improved resource allocation:** AI can help optimize resource allocation based on predicted timelines,

ensuring that we have the right resources at the right time.

- **Data-driven decision-making:** AI provides data-driven insights that support informed decision-making in project management, enhancing efficiency and project outcomes.

6. **Building a Data-Driven Culture:**

To fully leverage the power of AI in project timeline prediction, we need to build a data-driven culture. This involves:

- **Data collection and management:** Establishing systems for collecting and managing project data consistently and accurately.
- **Data analysis and interpretation:** Developing the skills and tools necessary to analyze and interpret project data effectively.
- **Integration with project management processes:** Integrating AI-powered tools into existing project management workflows.
- **Continuous improvement:** Regularly reviewing and refining AI models based on new data and project outcomes.

IDENTIFYING AND ASSESSING PROJECT RISKS

Imagine a world where project risks aren't just identified but actively anticipated, their likelihood and impact assessed with precision, and mitigation strategies prioritized before they even materialize. This is the reality AI is bringing to project management, revolutionizing the way we approach risk management.

AI's ability to analyze vast datasets, identify patterns, and make predictions based on historical data empowers project managers to gain a proactive edge in risk mitigation. Instead of reacting to unforeseen events, AI helps us anticipate, prioritize, and proactively address potential issues.

The Power of AI for Risk Identification

AI-powered tools are equipped with advanced algorithms capable of analyzing massive datasets encompassing historical project data, market trends, industry benchmarks, and even external factors like weather patterns and economic indicators. By analyzing this complex information, AI can identify potential risks that might otherwise go unnoticed.

Consider a construction project. AI can analyze historical data on past projects, identifying common risks associated with specific weather conditions, supplier delays, or material shortages. By analyzing this information, AI can proactively flag potential issues and recommend strategies for mitigating them.

AI-Driven Risk Assessment

AI doesn't just identify risks; it delves deeper to assess their likelihood and potential impact. Using machine learning algorithms, AI can evaluate the probability of each risk occurring based on historical data and current conditions. It can also assess the potential consequences of each risk, including financial loss, schedule delays, and safety hazards.

For example, an AI-powered risk assessment tool can evaluate the likelihood of a supplier delay based on the supplier's past performance, current market conditions, and potential disruptions to the supply chain. The tool can also assess the potential impact of this delay on project timelines, budgets, and overall project success.

Prioritizing Mitigation Efforts

Once risks are identified and assessed, AI helps prioritize mitigation efforts based on their likelihood and impact. The AI-driven risk assessment system can rank risks according to their severity and recommend strategies for minimizing their potential effects. This enables project managers to allocate resources effectively, focusing on the risks that pose the greatest threat to project success.

For instance, if an AI-powered tool identifies a high likelihood of a major supplier delay, it can recommend strategies for diversifying suppliers, securing alternative materials, and creating contingency plans to minimize the impact of the delay.

Benefits of AI-Driven Risk Management

- **Increased Accuracy and Efficiency:** AI eliminates human bias and error, providing more accurate risk assessments and enabling more efficient risk management practices.
- **Early Risk Detection:** AI can identify potential risks early in the project lifecycle, giving project managers ample time to develop mitigation strategies and prevent issues from escalating.
- **Proactive Risk Mitigation:** AI empowers project managers to move from a reactive to a proactive

approach to risk management, minimizing the impact of potential issues.

- **Data-Driven Insights:** AI provides data-driven insights into risk factors, enabling informed decision-making and helping project managers make strategic choices.
- **Resource Optimization:** By prioritizing risks, AI helps project managers allocate resources effectively, focusing on the areas that require the most attention.

The Future of AI-Driven Risk Management

As AI technology continues to evolve, we can expect even more sophisticated applications in risk management. Advanced AI models will be able to analyze real-time data, predict future events with greater accuracy, and provide even more personalized risk mitigation strategies.

AI is poised to transform project risk management, enabling project teams to identify, assess, and mitigate risks more effectively than ever before. By harnessing the power of AI, project managers can make informed decisions, reduce uncertainty, and improve the likelihood of project success.

Optimizing Resource Allocation

Imagine a scenario where you're managing a complex construction project with tight deadlines and a multitude of moving parts. You need to allocate resources effectively, balancing skilled labor, equipment availability, and project demands. This is where the power of artificial intelligence (AI) comes into play. AI algorithms can analyze vast amounts of data, including historical project data, current resource availability, and projected demands, to optimize resource allocation in a way that would be impossible for humans alone.

AI-powered resource allocation tools go beyond simple scheduling. They can consider factors like skill sets, experience levels, and even employee preferences to create balanced teams and maximize efficiency. By analyzing historical project data, AI can identify patterns and predict potential resource bottlenecks, allowing you to proactively adjust resource allocation and prevent delays. This predictive capability is essential for managing dynamic projects where unexpected changes can quickly impact resource needs.

Optimizing Resource Allocation: A Real-World Example

Consider a large-scale infrastructure project where you have a diverse workforce, including engineers, construction workers, and specialized technicians. AI-powered resource allocation tools can analyze the skill sets of each individual and match them to specific tasks, ensuring the right people are assigned to the right jobs. This not only increases efficiency but also optimizes the utilization of specialized skills, leading to better project outcomes.

For example, an AI system can analyze the experience levels of welders and identify those best suited for specific welding tasks, while simultaneously considering their availability and projected workload. This dynamic allocation ensures that skilled welders are not overloaded, while less experienced individuals can be assigned to simpler tasks, allowing them to learn and develop their skills.

Beyond individual skill sets, AI can also optimize resource allocation based on project demands. By analyzing the project schedule and anticipated workload, AI can identify potential bottlenecks and allocate resources accordingly. If a particular phase of the project requires a surge in manpower, the AI system can identify and deploy the necessary resources, preventing delays and ensuring a smooth workflow.

Benefits of AI for Resource Allocation

The implementation of AI in resource allocation offers several key benefits:

- **Increased efficiency:** By matching the right people to the right tasks and predicting potential bottlenecks, AI can streamline workflows and increase project efficiency.
- **Improved resource utilization:** AI algorithms can optimize resource allocation, ensuring that skilled personnel and equipment are used effectively, minimizing downtime and wasted resources.
- **Reduced costs:** By optimizing resource allocation, AI can help minimize labor costs, equipment rentals, and material waste, ultimately leading to reduced project costs.
- **Enhanced accuracy:** AI-powered analysis and prediction tools provide data-driven insights that can help improve the accuracy of resource allocation, minimizing risks and delays.
- **Improved decision-making:** AI provides project managers with real-time data and insights, empowering

them to make more informed decisions regarding resource allocation.

Challenges of Implementing AI for Resource Allocation

Despite its potential, implementing AI for resource allocation comes with its own set of challenges:

- **Data availability and quality:** AI algorithms rely on large amounts of high-quality data. If data is incomplete, inaccurate, or inconsistent, it can hinder the effectiveness of AI-powered resource allocation tools.
- **Integration with existing systems:** Integrating AI systems with existing project management software and databases can be complex and time-consuming.
- **Data security and privacy:** Handling sensitive data, such as employee information and project details, requires robust security measures and adherence to privacy regulations.
- **Resistance to change:** Some stakeholders might resist the adoption of AI-powered resource allocation tools due to concerns about job security or a lack of understanding about AI's capabilities.

Overcoming Challenges and Embracing the Future of Resource Allocation

To overcome these challenges, organizations need to invest in data quality, develop robust data security protocols, and provide training to their staff to ensure seamless AI integration. It's crucial to address concerns about job security and demonstrate how AI can enhance their roles, rather than replace them.

The future of resource allocation lies in the seamless integration of AI with human expertise. By leveraging the power of AI to analyze data, predict potential challenges, and optimize resource utilization, project managers can enhance efficiency, reduce risks, and achieve better project outcomes. It's time to embrace the transformative power of AI and embark on a new era of intelligent resource allocation in project management.

Cost Estimation and Budgeting with AI

Imagine a project manager armed with the power to predict costs with uncanny accuracy, anticipate potential overruns before they happen, and optimize budgets for maximum efficiency. This is the potential of AI in cost estimation and budgeting. By leveraging the ability of AI to analyze vast amounts of historical data and market trends, project managers can gain a level of financial foresight that was previously unimaginable.

Think about the traditional approach to cost estimation. It often relies on spreadsheets, manual calculations, and expert judgment, making it susceptible to human error and biases. AI, on the other hand, can analyze data from past projects, market fluctuations, and other relevant sources to generate highly accurate cost estimates.

Let's delve into how AI transforms cost estimation:

1. **Historical Data Analysis:**

AI algorithms are trained on massive datasets of historical project information, including past costs, materials prices, labor rates, and project durations. This historical data acts as a foundation for predicting future costs. AI can identify patterns and trends in historical data that may not be apparent to humans.

Example:

Consider a construction project where historical data reveals that concrete prices tend to rise during certain seasons. AI can analyze this data and factor it into future cost estimates, alerting the project manager to potential price fluctuations and helping them adjust their budget accordingly.

2. **Market Trend Analysis:**

AI algorithms can continuously monitor market trends, including commodity prices, labor costs, and inflation rates. This real-time analysis allows for dynamic cost estimations, reflecting the ever-changing economic landscape.

Example:

Imagine an IT project involving the development of a new software application. AI can monitor market trends for software development costs, identify potential price variations due to changing technology landscapes, and provide updated cost estimates to ensure the project budget remains aligned with market realities.

3. **Risk Identification and Mitigation:**

AI can analyze historical data to identify potential risks associated with cost overruns, such as supplier delays, material shortages, and unexpected changes in scope. By identifying these risks early on, project managers can proactively implement mitigation strategies to minimize their impact.

Example:

In an engineering project involving the construction of a bridge, AI can analyze historical data on similar projects and identify potential risks like weather delays or unforeseen geological challenges. This analysis allows the project manager to allocate contingency funds and develop alternative plans to minimize potential cost overruns.

4. Optimized Budgeting:

AI can use its cost estimations to optimize project budgets, allocating funds strategically to areas where they are most needed. AI can also recommend cost-saving measures based on its analysis of historical data and market trends.

Example:

Imagine a construction project where AI analyzes historical data and market trends to identify opportunities for cost savings, such as using alternative building materials or optimizing labor schedules. These recommendations allow the project manager to allocate funds more efficiently, ensuring that the project is completed on time and within budget.

5. Predictive Cost Overrun Detection:

AI can analyze project progress, actual costs incurred, and forecasted market trends to predict potential cost overruns. By alerting project managers to these potential overruns early on,

AI provides valuable time to adjust plans and mitigate the impact.

Example:

In an IT project involving the development of a mobile application, AI can monitor the project's progress, analyze actual costs, and identify potential cost overruns due to unforeseen delays or scope changes. This early warning allows the project manager to adjust the budget, reallocate resources, or revise the project timeline to avoid exceeding the allocated funds.

Beyond Traditional Budgeting

AI goes beyond traditional cost estimation and budgeting by introducing innovative approaches like:

1. **Scenario Analysis:**

AI can generate multiple cost scenarios based on different assumptions about market conditions, project risks, and potential changes in scope. This allows project managers to assess the financial impact of various decisions and make informed choices.

Example:

In a construction project involving the construction of a high-rise building, AI can create various cost scenarios based on different assumptions about material prices, labor availability, and potential delays. These scenarios help the project manager understand the financial implications of different project timelines and strategies.

2. Machine Learning for Price Prediction:

AI can leverage machine learning algorithms to predict future prices of materials, labor, and other project components. This predictive capability provides valuable insights for budgeting and financial planning.

Example:

Imagine an engineering project involving the construction of a power plant. AI can use machine learning to predict future prices of steel, concrete, and other key materials, enabling the project manager to make more accurate budget projections and ensure the project stays within financial constraints.

Real-World Applications: AI is already being used in numerous construction, IT, and engineering projects to revolutionize cost estimation and budgeting. Here are some real-world examples:

Construction

Cost Estimation Software: AI-powered software is being used to estimate costs for large-scale infrastructure projects, including bridges, tunnels, and high-rise buildings.

Construction Project Management Platforms: AI-driven platforms are analyzing historical data and market trends to provide accurate cost estimates and identify potential cost overruns, helping project managers make informed decisions and stay within budget.

Construction Material Procurement: AI is being used to analyze market trends and predict future material prices, enabling construction companies to procure materials at the best possible rates.

IT

Software Development Cost Estimation: AI algorithms are being used to estimate costs for software development projects, taking into account factors such as project complexity, team size, and technology stacks.

IT Project Management Software: AI-powered software is being integrated into IT project management platforms, providing data-driven insights for cost estimation, budget allocation, and risk mitigation.

Cloud Cost Optimization: AI is being used to analyze cloud computing usage and identify cost-saving opportunities, helping IT organizations optimize their cloud spend.

Engineering

Engineering Design Cost Estimation: AI is being used to estimate the cost of engineering designs, including factors like material usage, labor requirements, and construction time.

Engineering Project Management Systems: AI-powered engineering project management systems are providing data-driven insights for cost estimation, resource allocation, and risk assessment, helping engineers manage projects effectively.

Challenges and Considerations

While AI offers tremendous potential for cost estimation and budgeting, it is important to address potential challenges and considerations:

Data Quality: AI's accuracy relies heavily on the quality and completeness of the data used for training and analysis. Incomplete or inaccurate data can lead to biased or unreliable cost estimations.

Data Privacy: Managing data privacy is crucial, ensuring that sensitive financial and project information is handled responsibly and ethically.

Transparency and Explainability: AI algorithms can be complex, making it challenging to understand the reasoning behind their cost estimations. Transparency and explainability are crucial for building trust in AI-driven cost estimations.

Human Expertise: While AI can provide valuable insights, human expertise remains essential for understanding context, interpreting AI-generated results, and making informed decisions.

The Future of AI in Cost Estimation and Budgeting

The future of AI in cost estimation and budgeting is promising. As AI technology continues to advance, we can expect even more powerful and sophisticated applications, including:

Predictive Modeling: AI will become even more adept at

predicting future costs, taking into account a wider range of variables and scenarios.

Real-Time Cost Monitoring: AI will enable real-time monitoring of project costs, providing instant alerts for potential overruns and deviations from the budget.

Personalized Budgeting: AI will personalize budget recommendations based on specific project requirements, risk profiles, and market conditions.

Conclusion

AI is transforming the way project managers approach cost estimation and budgeting. By leveraging the power of AI, project managers can gain unprecedented financial foresight, make informed decisions, and optimize budgets for maximum efficiency. While challenges and considerations remain, the future of AI in cost estimation and budgeting holds immense potential for enhancing project success and achieving financial goals. As AI technology continues to evolve, we can anticipate even more innovative and transformative applications that will reshape the landscape of project management in construction, IT, and engineering.

Data Driven Decision Making in Project Management

Imagine a project manager facing a critical decision: allocate resources to a high-risk, high-reward task or focus on a more predictable but less impactful path. Traditionally, this decision would rely on gut feeling, experience, and perhaps some rudimentary analysis. However, the advent of AI brings a new dimension to this scenario, enabling data-driven decision-making that can transform project outcomes.

AI-powered analytics offers project managers a powerful tool for extracting meaningful insights from vast amounts of data. This data encompasses historical project data, market trends, real-time performance metrics, and external factors that influence project success. By analyzing this data, AI algorithms can identify patterns, predict future outcomes, and quantify the potential risks and rewards associated with different decisions.

Let's consider a real-world example. A construction company is bidding on a large-scale infrastructure project. The project manager needs to determine the optimal resource allocation and schedule to ensure on-time delivery and budget adherence. Using AI-powered analytics, the project manager can analyze historical data from similar projects, factoring in market conditions, material availability, and weather patterns. The AI system can then generate predictive models that forecast potential delays, identify potential bottlenecks, and recommend resource allocation strategies to mitigate risks.

This data-driven approach empowers the project manager to make more informed decisions, reducing reliance on intuition and increasing the likelihood of project success. AI can also help project managers identify hidden relationships and trends in data that might be overlooked through traditional analysis. This ability to uncover hidden patterns can lead to innovative solutions and strategies that optimize project outcomes.

However, the value of AI-driven insights goes beyond simply predicting outcomes; it lies in its ability to empower project managers to make proactive adjustments throughout the project lifecycle. AI can continuously monitor project progress, detect deviations from planned timelines, and provide early warning signs of potential risks. This real-time monitoring allows project managers to identify issues before they escalate and implement corrective measures to ensure project goals are met.

For instance, an IT project team is using AI-powered tools to track software development progress. The AI system analyzes code commits, bug reports, and test results to identify potential code quality issues and predict delivery delays. Based on this analysis, the project manager can adjust resource allocation, prioritize critical tasks, and communicate potential delays proactively to stakeholders.

AI also plays a crucial role in risk management, enabling project managers to assess potential risks with greater accuracy and develop more effective mitigation strategies. By analyzing historical data and identifying potential risk factors, AI can predict the likelihood and impact of various risks. This information allows project managers to prioritize their risk mitigation efforts, focusing on high-impact risks and developing tailored strategies to minimize their impact.

Consider a scenario where an engineering team is designing a complex bridge. AI-powered risk assessment tools can analyze historical data on similar bridge projects, identify potential seismic risks, and suggest design modifications to enhance the bridge's resilience. This proactive approach helps the project team mitigate potential risks and ensure the bridge's long-term stability.

Beyond predicting timelines and identifying risks, AI-powered analytics can also support project managers in optimizing resource allocation and maximizing project efficiency. By analyzing historical data on employee performance, skill sets, and task dependencies, AI algorithms can recommend optimal resource allocation strategies that balance skill requirements, workload distribution, and project deadlines. This enables project managers to assign tasks efficiently, maximize team productivity, and minimize unnecessary resource utilization.

Imagine a construction project where AI analyzes historical data on worker performance, weather patterns, and material availability to recommend the most efficient resource allocation for different phases of the project. This data-driven approach can help the project manager optimize labor deployment, minimize idle time, and ensure on-time completion.

Furthermore, AI can help project managers gain a deeper understanding of the factors that contribute to project success or failure. By analyzing data from completed projects, AI can identify key performance indicators (KPIs) that correlate with positive outcomes and pinpoint areas where improvements are needed. This enables project managers to refine their processes, improve team performance, and achieve better results on future projects.

However, the potential of AI in project management extends beyond simply analyzing data; it lies in its ability to create a virtuous cycle of continuous improvement. By providing data-driven insights and recommendations, AI empowers project managers to make informed decisions, implement changes, and track the impact of those changes on project performance. This iterative process of data analysis, decision-making, and performance monitoring creates a feedback loop that continuously improves project outcomes.

Let's consider an example of how AI can help a company implement continuous improvement in its project management processes. An engineering firm is using AI-powered tools to analyze project completion data and identify common bottle-necks and areas for improvement. Based on these insights, the project manager implements changes to streamline workflows, enhance communication channels, and improve resource allocation. The AI system then monitors the impact of these changes, providing real-time feedback and suggesting further adjustments to optimize project performance.

In conclusion, data-driven decision-making is a powerful force that can transform project management practices. AI-powered analytics provides project managers with the tools to analyze vast amounts of data, extract meaningful insights, and make more informed decisions throughout the project lifecycle. By integrating AI into project management processes, organizations can achieve greater efficiency, reduce risk, improve productivity, and enhance project outcomes. As AI continues to evolve, its role in project management will undoubtedly become even more trans-formative, empowering project managers to navigate the complexities of modern projects with unprecedented agility and effectiveness.

CHAPTER 6

AI FOR AUTOMATION IN PROJECT MANAGEMENT

AUTOMATING REPETITIVE TASKS

In the bustling realm of project management, where efficiency and precision reign supreme, the relentless repetition of certain tasks can often bog down progress and drain valuable time. Imagine a project manager drowning in a sea of data entry, meticulously crafting schedule updates, and painstakingly generating reports, all while juggling the complexities of project execution. This is where AI steps in, offering a revolutionary solution to liberate project managers from the clutches of routine tasks.

AI's ability to automate repetitive tasks is a game-changer for project management, particularly in engineering, IT, and construction. It empowers project managers to focus on strategic decision-making, fostering innovation and driving projects towards successful completion.

Let's delve into the transformative power of AI automation in project management, exploring how it can revolutionize various aspects of project execution, from data entry to reporting and document generation.

Data Entry: A Symphony of Automation

The tedious task of data entry is a ubiquitous challenge in project management. Imagine the sheer volume of data that project managers must meticulously input into spreadsheets and databases, from resource allocation and project timelines to budget tracking and task progress. This repetitive and error-prone process can consume hours of valuable time, hindering the flow of information and decision-making.

AI emerges as a beacon of efficiency, automating data entry tasks with remarkable precision. AI-powered tools can extract data from diverse sources, such as emails, invoices, and project documents, and seamlessly populate them into databases and spreadsheets. These tools are adept at recognizing patterns and extracting relevant information, eliminating the need for manual input and reducing the risk of human error.

Scenario: Construction Project Management

Consider a construction project manager overseeing a sprawling construction site. Every day, the manager receives a flood of invoices, material orders, and inspection reports. Manually entering this data into spreadsheets would be a time-consuming and error-prone task. However, with AI-powered data entry tools, the manager can simply scan these documents, and the AI system will automatically extract key information, such as dates, amounts, and vendor details, populating them into the project database. This automated process saves valuable time and ensures accuracy, freeing the manager to focus on strategic project decisions.

Reporting: Unveiling Insights at a Glance

Project reports provide essential insights into project performance, tracking progress, identifying challenges, and communicating updates to stakeholders. However, generating these reports can be a laborious and time-consuming process, often involving manual data extraction, formatting, and analysis.

AI empowers project managers by automating the generation of comprehensive and insightful reports, freeing up valuable time for analysis and decision-making. AI-powered reporting tools can analyze vast amounts of data, identify trends, and generate customized reports that provide a clear and concise picture of project performance. They can automatically generate dashboards, charts, and graphs, visualizing key metrics and enabling project managers to quickly grasp project progress and potential risks.

Scenario: Engineering Project Management

An engineering project manager is tasked with reporting on the progress of a complex infrastructure project, involving numerous subcontractors, diverse tasks, and a strict budget. Manually collecting, analyzing, and formatting data for a comprehensive project report would be a time-consuming endeavor. With AI-powered reporting tools, the manager can access a real-time dashboard that automatically aggregates data from various sources, such as project progress trackers, resource allocation databases, and budget spreadsheets. The AI system can then generate customized reports, highlighting key performance indicators, potential risks, and areas requiring attention. This automation streamlines the reporting process, providing the manager with valuable insights to guide project decisions.

Document Generation: Streamlining Communication

In the world of project management, communication is paramount. Project managers must consistently generate documents, such as proposals, contracts, progress reports, and meeting minutes. Creating these documents often involves repetitive tasks, such as formatting, copying and pasting data, and incorporating standard clauses.

AI can automate the generation of these documents, streamlining communication and saving valuable time. AI-powered document generation tools can leverage pre-defined templates and data sources to create customized documents. They can automatically populate documents with relevant information, apply formatting rules, and generate tables and charts, ensuring consistency and accuracy.

Scenario: Construction Project Bidding

A construction company is preparing a bid for a large-scale infrastructure project. The bidding process involves creating detailed proposals, incorporating project specifications, and outlining the company's qualifications. Manually compiling this information into a proposal would be a time-consuming process. With AI-powered document generation tools, the company can leverage pre-defined proposal templates and automatically populate them with data from project specifications, company profiles, and relevant pricing information. This automation streamlines the bidding process, ensuring consistent formatting and accurate data, enabling the company to submit bids efficiently and effectively.

Benefits of AI Automation for Project Managers

The benefits of AI automation in project management are numerous and far-reaching:

- **Increased Efficiency:** Automating tasks frees up project managers' time to focus on strategic decision-making, problem-solving, and collaboration.
- **Enhanced Productivity:** AI-powered workflow optimization leads to faster project completion, improved resource utilization, and increased overall productivity.
- **Reduced Errors:** AI minimizes human error by automating repetitive tasks and providing data-driven insights, ensuring greater accuracy and reliability.
- **Improved Decision-Making:** AI provides data-driven insights, supporting informed decision-making in areas such as risk assessment, resource allocation, and project prioritization.
- **Enhanced Collaboration:** AI can facilitate seamless communication and collaboration between project stakeholders, breaking down silos and ensuring everyone is aligned.

Best Practices for Implementing AI Automation

While the benefits of AI automation are undeniable, successful implementation requires careful planning and execution. Here are some best practices to ensure a smooth transition and optimize the benefits of AI:

- **Identify Suitable Tasks:** Start by identifying repetitive and time-consuming tasks that are ideal for automation. Focus on tasks that can be easily defined and have consistent inputs and outputs.
- **Choose the Right Tools:** Select AI tools and platforms that are tailored to your specific project requirements, industry, and budget.
- **Data Preparation and Validation:** Ensure that the data used to train AI algorithms is accurate, relevant, and clean. Validate the data regularly to ensure its accuracy and maintain model performance.
- **Pilot Projects and Gradual Rollout:** Begin with pilot projects to test the feasibility and effectiveness of AI automation before implementing it across the organization. This allows for adjustments and refinements before full-scale implementation.
- **Continuous Monitoring and Optimization:** Monitor the performance of AI-powered systems regularly and make adjustments to improve accuracy, efficiency, and effectiveness.

Conclusion: Embracing AI for a Smarter Project Management Future

In the dynamic world of project management, AI is no longer a futuristic concept; it's a powerful tool at our fingertips. By automating repetitive tasks, providing insightful data-driven recommendations, and fostering collaboration, AI empowers project managers to navigate the complexities of project execution, optimize resource allocation, and drive projects towards successful completion.

As AI continues to evolve, we can expect even more transformative applications in project management. AI-powered tools will become increasingly sophisticated, offering more comprehensive insights, automating more complex tasks, and facilitating seamless collaboration between project stakeholders. Embracing AI is not about replacing human expertise; it's about augmenting it, creating a smarter and more efficient project management future.

AI Powered Task Management

Imagine a world where project tasks magically prioritize themselves, deadlines adjust dynamically based on real-time progress, and resources are allocated with pinpoint accuracy. This isn't science fiction; it's the reality AI-powered task management is bringing to project management. AI is transforming how we approach tasks, moving beyond mere automation to a realm of intelligent assistance, making project management more efficient, effective, and ultimately, successful.

One of the core strengths of AI in task management lies in its ability to analyze vast amounts of data and identify patterns that would be invisible to human eyes. This data analysis isn't just about crunching numbers; it's about understanding the intricate relationships between tasks, resources, and deadlines. For example, AI can analyze historical project data to predict the likelihood of a task being delayed based on factors like team availability, complexity, and previous task performance. This predictive capability allows project managers to proactively address potential issues and allocate resources accordingly, minimizing delays and ensuring projects stay on track.

AI-Powered Task Prioritization

The ability to prioritize tasks effectively is paramount in project management. AI algorithms can learn from past project data and current project context to prioritize tasks based on several factors, including:

- **Urgency:** Tasks with critical deadlines or those that directly impact subsequent tasks receive higher priority.
- **Importance:** Tasks that contribute most significantly to the project's overall goals or those that have the highest impact on project success are given priority.
- **Dependencies:** Tasks with dependencies on other tasks are prioritized based on the urgency and importance of the dependent tasks.
- **Resource Availability:** AI algorithms can consider team member availability, skillsets, and workloads to optimize task assignments and avoid bottlenecks.

AI-Driven Task Assignment

AI goes beyond simply prioritizing tasks; it actively assists in assigning them to the most appropriate team members. By analyzing team member profiles, skillsets, and historical performance data, AI can make data-driven decisions about task allocation. This ensures that tasks are assigned to individuals who possess the necessary skills and experience to complete them effectively.

AI-powered task assignment systems can also consider factors like team member availability, current workload, and individual performance levels. This dynamic allocation ensures that workloads are balanced, avoids overburdening team members, and optimizes overall project efficiency.

Real-Time Task Progress Tracking

Traditionally, tracking task progress has been a manual and time-consuming process. AI changes the game by enabling real-time task progress tracking, providing project managers with constant insights into the status of each task. AI algorithms can analyze data from various sources, including task management tools, communication channels, and time tracking software, to create a holistic picture of task progress.

Real-time tracking is essential for identifying potential delays, bottlenecks, and resource constraints early on. This enables project managers to take corrective actions proactively and avoid costly delays. AI can also automatically generate progress reports and alerts, ensuring that stakeholders remain informed and that critical issues are addressed promptly.

AI's Impact on Project Team Dynamics

AI's role in task management extends beyond individual tasks; it also influences team dynamics. AI-powered systems can foster a more collaborative and efficient work environment by:

- **Streamlining Communication:** AI chatbots and virtual assistants can automate routine communication tasks, such as sending meeting reminders, answering frequently asked questions, and scheduling team meetings. This frees up team members to focus on more complex and strategic tasks.

- **Enhancing Collaboration:** AI-powered collaboration platforms can facilitate seamless communication and information sharing between team members, regardless of their location. This promotes team cohesiveness and ensures everyone is working from the same information.

- **Providing Real-Time Feedback:** AI can analyze team performance data to provide real-time feedback to individuals and teams. This helps identify areas for improvement, track progress, and foster a culture of continuous learning.

Transforming Project Management

AI-powered task management is not just about automating tasks; it's about empowering project managers with the tools they need to make smarter decisions, optimize workflows, and deliver projects successfully. By leveraging AI's capabilities in task prioritization, assignment, progress tracking, and team collaboration, project managers can transform their approach to project management, achieving greater efficiency, productivity, and project outcomes.

Real-World Examples of AI in Task Management

- **Construction Project Management:** AI is being used to optimize construction schedules, allocate resources efficiently, and predict potential delays. By analyzing historical data and considering factors like weather conditions, material availability, and team availability, AI can create more accurate schedules and minimize disruptions.

- **Software Development:** AI-powered task management systems are helping software development teams manage complex projects by automating tasks, prioritizing tasks based on urgency and complexity, and providing real-time progress tracking. This helps teams stay on schedule, identify potential bottlenecks, and deliver high-quality software.
- **IT Project Management:** AI is being used to automate routine tasks, such as data entry, reporting, and scheduling. This frees up IT project managers to focus on more strategic tasks, such as risk management, stakeholder communication, and project planning.

Challenges and Considerations for AI in Task Management

While AI holds immense potential for project management, its implementation comes with challenges that need careful consideration:

- **Data Quality and Bias:** The accuracy and reliability of AI-driven task management systems depend heavily on the quality and completeness of the data used to train the algorithms. Biased data can lead to biased decisions, so it's crucial to ensure that the data used is accurate, representative, and unbiased.

- **Transparency and Explainability:** Understanding how AI algorithms make decisions is crucial for building trust and ensuring accountability. Explainable AI systems are being developed to provide insights into the reasoning behind AI decisions, enhancing transparency and user confidence.
- **Human-AI Collaboration:** AI is a powerful tool, but it's not meant to replace human judgment. Effective AI implementation requires a collaborative approach, where AI provides insights and recommendations, and human project managers utilize their experience and judgment to make informed decisions.

The Future of AI in Task Management

As AI technology continues to evolve, we can expect even more transformative applications in task management. The development of advanced AI algorithms, coupled with the increasing availability of data, will further enhance AI's ability to understand complex project dynamics, optimize resource allocation, and predict project outcomes.

The future of project management lies in embracing AI's power to optimize workflows, enhance collaboration, and ultimately, deliver projects more effectively and efficiently. By combining human expertise with AI's intelligent assistance, we can unlock a new era of project management characterized by greater productivity, innovation, and success.

This is a new frontier in project management, where AI acts as a powerful ally, supporting project managers in navigating complex challenges and achieving extraordinary results. The journey towards AI-powered task management is just beginning, and the possibilities are truly endless.

CHAPTER 7

———————

AI for Decision Making in Project Management

AI as a Decision Making Partner for Project Managers

Imagine a world where project managers are not burdened by mountains of data, where decisions are not based on gut feelings but on insights gleaned from vast amounts of information, and where risks are proactively identified and mitigated before they derail projects. This is the promise of artificial intelligence (AI) in project management, transforming the way projects are conceived, planned, executed, and delivered.

AI is not a replacement for project managers; rather, it acts as a powerful decision-making partner, augmenting their expertise and empowering them with data-driven insights. It's like having a sophisticated, always-on analyst at your fingertips, providing real-time information, predictions, and recommendations to guide your every move.

Consider the scenario of a large-scale construction project, where every decision carries significant financial and time implications. AI can analyze historical data from similar projects, weather patterns, and resource availability to predict potential delays, identify risks, and suggest proactive mitigation strategies.

The value of AI as a decision-making partner goes beyond individual tasks; it extends to strategic decision-making for the entire project. By providing data-driven insights, AI empowers project managers to make informed decisions about project scope, budget, and timeline, ensuring alignment with business goals and maximizing the chances of project success.

How AI Augments Human Decision-Making

It's important to understand that AI is not a replacement for human judgment; rather, it's a powerful tool that enhances decision-making by providing:

- **Data-Driven Insights:** AI can analyze vast amounts of data from various sources to identify patterns, trends, and insights that might be missed by humans. This data-driven perspective helps project managers make more informed and accurate decisions.
- **Predictive Analysis:** AI algorithms can use historical data to predict future outcomes, such as project timelines, risk probabilities, and cost overruns. This allows project managers to anticipate challenges and proactively address them before they become major problems.
- **Scenario Planning:** AI can generate multiple scenarios based on different assumptions and variables, allowing project managers to evaluate the potential consequences of various decisions. This helps them

identify the most effective strategies for achieving project goals.

- **Risk Mitigation:** AI can identify potential risks, assess their likelihood and impact, and recommend mitigation strategies. This proactive approach reduces the chances of project derailment and ensures a smoother project journey.

- **Resource Optimization:** AI can analyze project requirements, team skills, and resource availability to recommend optimal resource allocation. This ensures that the right people with the right skills are assigned to the right tasks, maximizing team productivity and project efficiency.

Embracing AI for a Brighter Future

The future of project management is undeniably intertwined with AI. By embracing AI as a decision-making partner, project managers can unlock a new era of efficiency, productivity, and innovation. The journey will not be without its challenges, but the rewards of AI-driven project management are undeniable.

By addressing data quality issues, prioritizing data privacy and security, investing in AI expertise, and upholding ethical principles, organizations can leverage the transformative power of AI to enhance their project management capabilities, elevate project outcomes, and stay ahead in the competitive landscape.

CHAPTER 8

AI for Collaboration and Communication in Project Management

AI Powered Communication Tools

Imagine a world where project teams can effortlessly communicate, collaborate, and stay informed in real time, all thanks to the power of artificial intelligence (AI). This is the reality that AI-powered communication tools are bringing to project management, revolutionizing how teams interact and exchange information.

One of the most prominent examples of AI-powered communication tools is **chatbots**, intelligent programs that can engage in human-like conversations. Imagine a chatbot integrated into your project management software, acting as a virtual assistant that can answer your questions, provide project updates, and even resolve simple issues without human intervention. This frees up project managers and team members to focus on more complex tasks, streamlining communication and improving efficiency.

For instance, a chatbot could be programmed to:

- **Answer basic questions about project timelines, deliverables, and team members.** This eliminates the need for constant email inquiries, freeing up time for more strategic communication.
- **Provide real-time project updates to stakeholders.** Chatbots can automatically update stakeholders on project progress, milestones, and any changes or delays, ensuring transparency and keeping everyone on the same page.
- **Handle routine tasks like scheduling meetings, booking resources, and sending reminders.** By automating these mundane tasks, chatbots free up team members to focus on their core responsibilities.

Another game-changer in AI-powered communication is **virtual assistants**, software applications that can perform various tasks based on voice commands or text instructions. Virtual assistants can be integrated into project management platforms, enabling team members to:

- **Dictate notes, create tasks, and set reminders without manually typing.** This is especially useful for mobile-first teams or those working in fast-paced environments where efficiency is paramount.
- **Access and share information quickly and easily.** Virtual assistants can search project databases, retrieve relevant documents, and share information with other team members instantly.
- **Manage meeting schedules and coordinate team calendars.** Virtual assistants can automatically

schedule meetings based on team availability, send meeting reminders, and even generate meeting agendas.

Beyond chatbots and virtual assistants, **automated meeting scheduling** is revolutionizing the way project teams plan and execute meetings. This technology leverages AI algorithms to analyze team availability, meeting objectives, and preferred meeting times, automatically scheduling meetings that optimize efficiency and collaboration.

Imagine the benefits:

- **No more back-and-forth emails or phone calls to find a time that works for everyone.** The AI-powered scheduler handles the entire process, saving valuable time for team members.
- **Meetings are more productive, as they are scheduled at times that are convenient for all participants.** This increases engagement and maximizes the effectiveness of meetings.
- **Meetings are automatically scheduled based on project milestones, ensuring that important discussions are prioritized and scheduled at the right time.** This ensures that project deadlines are met and progress remains on track.

AI-powered communication tools are also transforming how teams collaborate on documents and share knowledge. **AI-powered document management platforms** leverage machine learning to automatically organize, categorize, and index documents, making it easier for teams to find the information they need. These platforms can also provide insights into document usage and identify areas where knowledge sharing could be improved.

Furthermore, AI is being incorporated into **knowledge management systems**, enabling teams to easily capture, share, and access knowledge across the entire project lifecycle. This ensures that valuable insights and lessons learned from past projects are readily available for future projects, fostering continuous improvement and innovation.

Here are some additional ways AI is enhancing collaboration and communication in project management:

- **AI-powered translation tools:** Enabling teams to communicate effectively across language barriers, breaking down communication silos and fostering international collaboration.
- **Sentiment analysis tools:** Analyzing communication patterns and sentiment to identify potential conflicts or misunderstandings, allowing project managers to intervene proactively and maintain positive team dynamics.
- **AI-powered collaboration platforms:** Providing a central hub for communication, task management, and document sharing, fostering seamless collaboration and enhancing team productivity.

The integration of AI into communication and collaboration tools is transforming the project management landscape. By automating routine tasks, providing real-time information sharing, and facilitating seamless collaboration, AI empowers teams to communicate effectively, work efficiently, and achieve project success. As AI continues to evolve, we can expect even more innovative tools and platforms that will further enhance collaboration and communication in project management, paving the way for more efficient, productive, and successful projects in the future.

Improving Team Collaboration and Communication

AI is revolutionizing communication and collaboration in project management by enabling seamless information flow, fostering better communication practices, and enhancing team productivity. Here's how:

- **Real-Time Information Sharing and Transparency:** AI enables real-time information sharing, ensuring that everyone on the team has access to the latest updates, project documents, and key data. This transparency fosters trust and collaboration, reducing the risk of miscommunication and delays.
- **AI-Powered Knowledge Management:** AI-powered knowledge management platforms can help project teams share best practices, lessons learned, and relevant information. This central repository of knowledge can help teams improve their processes, avoid repeating mistakes, and learn from each other.
- **Collaborative Decision-Making:** AI can facilitate collaborative decision-making by analyzing data, identifying potential risks and opportunities, and suggesting alternative solutions. This data-driven approach helps teams make more informed decisions, leading to better outcomes.

- **Virtual Team Collaboration:** AI plays a critical role in enabling virtual team collaboration, which is becoming increasingly common in today's interconnected world. AI-powered tools like video conferencing platforms with real-time translation features allow teams to communicate seamlessly, regardless of physical location.

Building a Collaborative and AI-Enabled Project Environment

- **Encourage AI Literacy:** Educate project team members about AI, its capabilities, and limitations. This will help them understand how AI can support their work and encourage them to embrace AI-powered tools.
- **Foster a Culture of Collaboration:** Create a work environment that values teamwork, open communication, and shared learning. This will help teams effectively leverage AI for communication and decision-making.
- **Start Small and Scale Gradually:** Instead of trying to implement AI across all aspects of project management at once, start with small-scale pilots and gradually expand the use of AI as the team gains experience and confidence.

By embracing AI-powered communication and collaboration tools, project teams can overcome traditional barriers, work more efficiently, and achieve better project outcomes. As AI technology continues to evolve, we can expect even more innovative solutions that will transform the way project teams interact, communicate, and collaborate in the future.

Conclusion

AI is revolutionizing project management by enabling real-time information sharing and transparency. This fosters collaboration, empowers better decision-making, and improves project outcomes. However, it is crucial to address the challenges associated with AI implementation, including data security, accuracy, and bias, to maximize the benefits of this transformative technology. By embracing AI-powered solutions for transparency, project managers can create a more collaborative, efficient, and successful project environment.

AI Powered Document Management and Knowledge Sharing

In the ever-evolving landscape of project management, seamless information access and knowledge sharing are paramount to project success. AI-powered document management systems offer a revolutionary approach to tackling the challenges associated with managing vast amounts of project documentation and ensuring that vital information is readily available to all stakeholders.

Imagine a project team where every document, from design specifications to meeting minutes, is instantly searchable and accessible. AI-powered document management systems make this a reality by leveraging natural language processing (NLP) and machine learning algorithms to analyze and index documents with unmatched accuracy. This eliminates the time-consuming manual search processes that often hinder project progress.

One of the most significant advantages of AI in document management lies in its ability to extract key insights from complex documents. By applying machine learning models trained on vast datasets of project documentation, these systems can automatically identify relevant keywords, concepts, and relationships within documents. This enables project teams to quickly grasp the core essence of any document, even if it spans hundreds of pages.

Take, for instance, a construction project involving a complex network of subcontractors. AI-powered document management can analyze contracts, specifications, and communication logs to identify potential conflicts or delays before they arise. By proactively highlighting areas of concern, AI can empower project managers to take preventative measures and avoid costly setbacks.

Beyond document analysis, AI plays a critical role in knowledge sharing and collaboration within project teams. These systems can automatically categorize documents based on content, project phase, or stakeholder relevance, making it easy for team members to locate relevant information quickly.

Consider a scenario where a project team is developing a new software application. An AI-powered system can automatically create a knowledge base by indexing all project-related documents, code repositories, and communication records. This knowledge base becomes a centralized hub for information, allowing team members to access critical insights, learn from past experiences, and avoid duplicating efforts.

Moreover, AI can enhance the collaborative aspects of document management by facilitating real-time document co-authoring and version control. This empowers team members to work simultaneously on documents while ensuring that all changes are tracked and coordinated, minimizing confusion and streamlining collaboration.

The benefits of AI-powered document management and knowledge sharing extend beyond efficiency and collaboration. These systems can also contribute to a more transparent and accountable project environment.

Imagine a scenario where a client requires detailed information on a specific project phase. With AI-powered document management, project managers can instantly provide a comprehensive overview, including relevant documents, communication logs, and progress reports. This level of transparency fosters trust and enhances stakeholder satisfaction.

However, as with any transformative technology, the successful implementation of AI in document management requires careful planning and consideration. It is crucial to ensure that data privacy and security protocols are robustly in place to protect sensitive project information.

Furthermore, the chosen AI platform must be compatible with existing project management systems and workflows to avoid disrupting existing practices. Training team members to effectively utilize the AI system is essential to unlock its full potential and ensure widespread adoption.

As project management continues to embrace digital transformation, AI-powered document management and knowledge sharing will play an increasingly vital role in driving efficiency, collaboration, and project success. By leveraging the power of AI to streamline information access and empower project teams with actionable insights, organizations can unlock the full potential of their projects and navigate the complexities of today's fast-paced and competitive environment.

CHAPTER 9

CHALLENGES AND FUTURE OF AI IN PROJECT MANAGEMENT

CHALLENGES OF AI INTEGRATION IN PROJECT MANAGEMENT

The integration of AI into project management, while promising numerous benefits, presents significant challenges that need to be addressed for successful implementation. These challenges encompass various aspects, from technical hurdles to organizational and societal implications.

TECHNICAL CHALLENGES:

Data Quality and Availability: AI algorithms are heavily reliant on high-quality and sufficient data. Project management often involves diverse data sources, some of which may be incomplete, inconsistent, or outdated. The availability and quality of data can significantly impact the accuracy and effectiveness of AI-driven insights. For example, using AI to predict project timelines requires accurate historical data on past projects, including factors like resource availability, task dependencies, and external influences. Incomplete or inaccurate data can lead to misleading predictions, hindering effective decision-making.

Integration with Existing Systems: Integrating AI tools and platforms with existing project management software and workflows can be complex and time-consuming. Different software systems may have incompatible data formats, APIs, or security protocols, requiring significant effort to bridge the gaps and ensure seamless data flow. For example, integrating an AI-powered risk assessment tool with a traditional project management software may require custom integrations to synchronize data and ensure consistent updates.

Algorithm Complexity and Black Box Problem: While AI algorithms can provide valuable insights, their inner workings can be complex and opaque, leading to the "black box problem." Understanding how AI arrives at specific conclusions can be challenging, making it difficult to interpret results, assess bias, or troubleshoot issues. This lack of transparency can be a major barrier to trust and adoption, especially in industries where decision-making requires accountability and explainability. For instance, when an AI-powered scheduling tool identifies a potential delay, it may be difficult to understand the underlying reasoning and identify the specific factors contributing to the prediction. This lack of explainability can make it challenging for project managers to validate the AI's recommendations and make informed decisions.

Organizational Challenges:

Change Management and Resistance to AI: Introducing AI into established project management practices can face resistance from team members who may be hesitant to adopt new technologies or fear job displacement. Effective change management strategies are crucial to address concerns, provide training, and foster a culture of innovation and collaboration around AI adoption. For instance, introducing an AI-powered task management tool may require training for project team members to understand its functionalities, interpret its recommendations, and utilize its features effectively. Effective change management can involve demonstrating the benefits of AI, providing hands-on training, and addressing concerns about job security.

Skill Gaps and Training Needs: Implementing AI in project management requires skilled professionals with expertise in data science, AI algorithms, and project management best practices. Addressing the skill gap through training programs and recruitment strategies is crucial to building a workforce capable of managing AI-driven projects effectively. For example, organizations may need to invest in training programs to equip project managers with the necessary skills to understand AI algorithms, interpret data-driven insights, and leverage AI tools to optimize their projects. They may also need to recruit data scientists and AI specialists to support the development and implementation of AI solutions.

Cost and Investment Considerations: Implementing AI solutions can involve significant initial investments in software, hardware, data infrastructure, and training. Organizations need to carefully assess the cost-benefit analysis, considering the potential return on investment, the long-term impact on project outcomes, and the value of AI-driven efficiency and innovation. For example, implementing an AI-powered risk management system may require investment in data collection, analysis, and tool licensing. However, the potential benefits of improved risk mitigation and reduced project delays can justify the investment in the long run.

Societal Challenges:

Ethical Considerations and Bias: AI algorithms can reflect and amplify existing biases present in the data used to train them. This can lead to discriminatory outcomes or unfair advantages for certain groups. Addressing ethical considerations and mitigating bias in AI development and deployment is crucial to ensure fairness and equity in project management practices. For instance, using AI to analyze historical data for resource allocation may perpetuate existing biases, potentially favoring certain individuals or groups over others. It's essential to ensure that AI systems are trained on diverse and representative data and are evaluated for potential bias before deployment.

Data Privacy and Security: AI applications in project management often involve the collection and processing of sensitive data, raising concerns about data privacy and security. Implementing robust data security measures, complying with privacy regulations, and obtaining informed consent from individuals are essential to protect personal data and maintain trust. For example, when using AI-powered tools to analyze project performance data, organizations need to ensure that data is anonymized or pseudonymized to protect individual privacy. They also need to comply with relevant data protection regulations, such as the General Data Protection Regulation (GDPR), to maintain ethical data handling practices.

Navigating these challenges requires a multifaceted approach. Organizations need to invest in building data infrastructure, upskilling their workforce, and implementing robust governance frameworks for ethical and responsible AI development. By actively addressing these challenges, organizations can harness the transformative potential of AI to enhance project management efficiency, optimize project outcomes, and drive innovation in the engineering, IT, and construction industries.

Skilled Professionals for Managing AI Driven Projects

The successful implementation and management of AI-driven projects demand a new breed of professionals equipped with a unique blend of skills. These individuals must not only be adept at traditional project management methodologies but also possess a deep understanding of AI, data science, and the intricacies of its application within their respective industries.

At the core of this new generation of project professionals lies a deep understanding of AI. This encompasses a thorough knowledge of its fundamental principles, including machine learning, deep learning, and natural language processing. Individuals must be able to discern the capabilities of AI, understand its limitations, and apply it effectively to address specific project challenges.

Data science skills become indispensable in this context. AI thrives on data, and professionals must be proficient in data analysis, interpretation, and visualization. They need to know how to cleanse, prepare, and manage large datasets, ensuring their quality and reliability. Moreover, the ability to extract meaningful insights from these datasets, translate them into actionable strategies, and communicate them effectively to project stakeholders is crucial.

Furthermore, these professionals must seamlessly integrate AI into existing project management frameworks. This necessitates strong project management expertise, encompassing skills like planning, scheduling, resource allocation, risk management, and stakeholder communication. They must be able to leverage AI to enhance these traditional processes, optimize workflows, and achieve project objectives more efficiently.

Here's a breakdown of the key skill sets required to manage AI-driven projects.

AI Expertise:

- **Understanding AI Concepts:** A solid grasp of the fundamental principles of AI, including machine learning, deep learning, natural language processing, and computer vision.
- **AI Algorithm Selection:** Knowledge of various AI algorithms and the ability to select the most appropriate algorithm for specific project tasks.
- **Data Preprocessing:** Skills in data cleaning, transformation, and preparation for AI model training.
- **AI Model Training and Evaluation:** Experience in training and evaluating AI models, ensuring accuracy, reliability, and robustness.
- **AI Ethics and Bias:** Awareness of ethical considerations in AI development and deployment,

including bias mitigation and responsible AI practices.

Data Science Skills:

- **Data Analysis and Interpretation:** Ability to analyze large datasets, identify patterns and trends, and draw meaningful insights.
- **Statistical Modeling:** Knowledge of statistical methods and techniques for data analysis and prediction.
- **Data Visualization:** Skills in creating informative and compelling data visualizations to communicate insights to stakeholders.
- **Data Management:** Expertise in managing and organizing large datasets, ensuring data quality and integrity.
- **Big Data Analytics:** Understanding of big data technologies and techniques for analyzing massive datasets.

Project Management Expertise:

- **Project Planning and Scheduling:** Proficiency in developing project plans, timelines, and schedules.
- **Resource Allocation:** Skills in optimizing resource allocation, considering skills, availability, and project requirements.
- **Risk Management:** Knowledge of risk identification, assessment, mitigation, and contingency planning.
- **Stakeholder Management:** Ability to effectively

communicate with stakeholders, manage expectations, and resolve conflicts.

- **Agile Methodologies:** Understanding of agile project management frameworks and their integration with AI.

Domain-Specific Knowledge:

- **Engineering Project Management:** Understanding of engineering principles, construction processes, and design optimization.
- **IT Project Management:** Familiarity with software development methodologies, cybersecurity practices, and data management.
- **Construction Project Management:** Knowledge of construction scheduling, cost estimation, quality control, and safety protocols.

Collaboration and Communication Skills:

- **Effective Communication:** Ability to clearly communicate technical concepts to non-technical stakeholders.
- **Team Collaboration:** Skills in leading and working effectively within multi-disciplinary teams.
- **Cross-Functional Communication:** Expertise in collaborating with professionals from different disciplines, including AI specialists, data scientists, and project managers.

Key Soft Skills:

- **Critical Thinking:** The ability to analyze data, evaluate AI insights, and make informed decisions.
- **Problem Solving:** Skills in identifying problems, exploring solutions, and implementing effective strategies.
- **Communication:** Clear and concise communication skills to effectively convey complex information to stakeholders.
- **Teamwork:** The ability to collaborate effectively with diverse teams, fostering a culture of innovation and mutual support.
- **Adaptability:** Flexibility and willingness to learn new technologies and adapt to rapidly evolving AI landscapes.
- **Ethical Awareness:** Understanding the ethical implications of AI and ensuring its responsible development and deployment.

THE FUTURE OF AI IN PROJECT MANAGEMENT

The future of AI in project management is brimming with exciting possibilities, promising to reshape how projects are planned, executed, and delivered. Several emerging trends are poised to revolutionize the field, leading to enhanced efficiency, improved decision-making, and greater project success. Looking ahead, AI is poised to revolutionize project management practices in profound ways. Emerging technologies like generative AI, natural language processing, and machine learning will continue to enhance our ability to predict, plan, execute, and optimize projects. As AI continues to evolve, project management will become more data-driven, efficient, and sustainable. AI will play a crucial role in achieving project goals while minimizing environmental impact, fostering safer working environments, and promoting responsible resource utilization.

AI is not replacing project managers but transforming their roles. Project managers will need to adapt their skillsets to leverage AI effectively, becoming:

- **AI-Savvy Decision-Makers:** Project managers will need to understand how AI works, interpret its outputs, and make informed decisions based on data-driven insights.
- **Strategic Problem Solvers:** AI will automate routine tasks, allowing project managers to focus on more complex problem-solving and strategic planning.
- **Leaders in AI-Driven Teams:** Project managers will need to guide and mentor their teams on how to effectively use AI tools and technologies.

Call to Action: To navigate the exciting future of AI in project management, we must take a proactive approach:

- **Invest in AI education and training:** Equip yourself and your team with the knowledge and skills to effectively integrate AI into your projects.
- **Embrace AI tools and platforms:** Explore and implement AI-powered project management software, data analytics tools, and communication platforms.
- **Foster a data-driven culture:** Encourage data-driven decision-making and prioritize data quality and security within your organization.
- **Collaborate with AI experts:** Partner with data

scientists, AI engineers, and technology consultants to leverage their expertise in integrating AI solutions.

- **Stay informed about AI trends:** Continuously research and learn about emerging AI technologies and their potential impact on project management.

By embracing these actions, we can harness the power of AI to create a future where projects are executed with greater efficiency, sustainability, and innovation. The journey ahead is exciting, and with the right mindset and proactive approach, we can leverage AI to transform project management and achieve extraordinary results.

Acknowledgments

The completion of this book wouldn't have been possible without the unwavering support and guidance of several individuals.

First and foremost, I want to express my deepest gratitude to [Name of person or group], whose encouragement and belief in this project have been instrumental. Their insightful feedback and invaluable contributions have shaped the content and direction of this book.

Their expertise and willingness to share their knowledge have been invaluable.

Finally, I want to thank [Name of person or group] for their patience and understanding during the writing process. Their support has made this journey much smoother.

Afterword

The creation of "The N.E.R.D.Y. Way: An Everyday Guide to AI" was a collaborative effort, and we are deeply grateful to everyone who contributed their expertise, insights, and unwavering support.

First and foremost, we would like to express our sincere gratitude to the 3CAT team for their meticulous research, insightful contributions, and dedication to crafting engaging and accessible content. Their collective expertise and passion were instrumental in bringing this book to life.

We extend our heartfelt thanks to the reviewers who provided valuable feedback and guidance throughout the writing process. Their thoughtful suggestions and constructive criticism helped shape the book into its final form.

We are also grateful to the individuals and organizations who generously shared their knowledge and experience, contributing to the depth and accuracy of the book. Their insights have enriched the content and provided valuable context for our readers.

Finally, we would like to thank our families and friends for their patience and understanding during the long hours spent writing and editing this book. Their unwavering support has been a source of inspiration and strength.

This appendix provides additional resources and information to complement the content discussed in the book.

Glossary

This glossary provides definitions for key terms and concepts discussed in the book, making it easier for readers to navigate the complex world of AI.

1. **Artificial Intelligence (AI):** The simulation of human intelligence processes by computer systems, typically involving learning, problem-solving, and decision-making.
2. **Machine Learning (ML):** A subset of AI that enables computer systems to learn from data without explicit programming.
3. **Deep Learning (DL):** A type of ML that uses artificial neural networks with multiple layers to learn complex patterns from data.
4. **Natural Language Processing (NLP):** A branch of AI that enables computers to understand, interpret, and generate human language.
5. **Predictive Analytics:** The use of statistical techniques to analyze historical data and predict future outcomes.
6. **Automation:** The use of technology to automate repetitive tasks and processes.
7. **Data Science:** The field of study that involves extracting knowledge and insights from data.
8. **Algorithms:** Sets of instructions that define the steps involved in solving a specific problem.
9. **Data Sets:** Collections of data that are used to train and evaluate AI models.

10. **Neural Networks:** Computational models inspired by the structure and function of the human brain.
11. **Project Management:** The process of planning, organizing, and managing resources to achieve specific project goals.

References & Sources

This section provides a list of references and sources that were consulted in the writing of this book.

1. Cowell, Alan. "Overlooked No More: Alan Turing", Condemned Code Breaker and Computer Visionary. The New York Times. June 5, 2019.

2. Kanabar, Vijay and Jason Wong. *The AI Revolution in Project Management: Elevating Productivity with Generative AI*, 1st Edition, Pearson, 2024.

3. Boudreau , Paul. *Applying Artificial Intelligence In Project Management* Kindle Edition.

4. Matthews , Bradon, "AI in Project Management: Uses, Impacts, & 2024 Trends", Published September 27, 2024. https://project-management.com/ai-project-management/.

5. Shaping the Future of Project Management With AI, Project Management Institute, October 2023. https://www.pmi.org/learning/thought-leadership/ai-impact/shaping-the-future-of-project-management-with-ai.

6. Pratt, Mary K. "How AI is transforming project management", March 7, 2024, TechTarget, https://www.techtarget.com/searchenterpriseai/feature/How-AI-is-transforming-project-management#:~:text=How%20AI%20is%20used%20in,available%20resources%20and%20other%20data.

7. Shehab, Faisal, "How AI Can Revolutionize Project Management", LinkedIn, July 17, 2023, https://www.linkedin.com/pulse/how-ai-can-revolutionize-project-management-dr-faisal-shehab.

8. Nieto-Rodriguez, Antonio and Ricardo Viana Vargas, "How AI Will Transform Project Management", Harvard Business Review, February 2, 2023, https://hbr.org/2023/02/how-ai-will-transform-project-management.

ABOUT THE AUTHOR
DR. C.B. HOWARD, 3CAT

3CAT is a collective of professionals collaborating across various disciplines to offer innovative and practical solutions for individuals seeking to comprehend the effects of artificial intelligence (AI). We are committed to the dissemination of education and information, striving to enhance the lives of others. While knowledge is a powerful tool, its true potential is realized through its application.

At 3CAT, we acknowledge that everyone is unique. Consequently, we provide a diverse range of training and guidance materials tailored to accommodate different needs and learning preferences. Our publications cover a wide range of important topics, aiming to deepen understanding and knowledge in various areas of interest.

For every publication, 3CAT collaborates to conduct research, develop pertinent topics, create manuscripts, and oversee the publication process, to ensure the highest quality work possible.

The **N.E.R.D.Y.** WAY is an acronym for k**N**owledge, **E**ducation, **R**esource, **D**iscovery for **Y**ou.

The trademark for "The NERDY WAY" has been applied for and is currently pending.